I0005955

WORKING DRAWINGS

BY

ARTHUR B. BABBITT

TEACHER OF MECHANICAL DRAWING, MANUAL TRAINING
DEPARTMENT, HARTFORD, CONN., PUBLIC HIGH SCHOOL

NEW YORK
HENRY HOLT AND COMPANY
1911

PREFACE

This book is designed to cover two different fields. As an elementary text-book it is complete in itself and should enable the student to make or read any simple mechanical drawing. As an introductory work, in schools that offer a three or four years' course in Mechanical Drawing, it should lay a good foundation for the subject of Projection.

The drawing instruments are explained as they are introduced in the course, instead of in a chapter by themselves. Each problem is fully discussed before the student is expected to attack it. It is suggested that the student be required to work out his problem for each plate, free-hand, outside of the class-room. This will give him valuable practice in free-hand sketching, and it will save time in the class-room for needed attention to technique.

The exercises in lettering (Chapter XIV), should show results in the improved appearance of all drawings made. The Geometrical definitions are usually wanting in books on Mechanical Drawing; but are inserted here (Chapter XIV). Even if not set as a lesson to be learned, they will still be found very useful for reference. The geometrical exercises in Chapter XVII are accompanied either by a reference to the solution of the problem given in Chapter XVI, or by the note "original," signifying that the student is to work out the problem for himself.

The course may be given without home-work, but the author is a firm believer in home-work. Where such work is required, the results immediately aimed at **are better,** and in many indirect ways the subject is found to have a **higher educational** value. The author has, therefore, been rather liberal in suggestions for outside work, and has added extra plates for the special benefit of the ambitious student.

The book represents the work of a year, two forty-five minute periods a week, and is the result of a ten years' testing process in the class-room, with high school students in the regular course, and with young machinists in evening classes.

A. B. B.

Hartford, Conn.
  Aug. 15, 1911.

CONTENTS

INDEX TO PLATES

CHAPTER I

MATERIAL

An elaborate equipment is not necessary for good work in mechanical drawing, but serviceable material which with careful handling will produce accurate results is required. The list given below covers all that is needed to solve almost any problem, and is illustrated fully in Fig. 1.

Drawing-board	Ink and pencil erasers
T square	Pencil-sharpener
Triangle, 30°–60°	Thumb tacks
Triangle, 45°	Bottles of drawing ink, red
Scale	and black
Scroll	Penholder, writing pens
Pencils, 3H and 6H	Set of drafting instruments

Drawing-board—The drawing-board should be made of soft pine with the left or working edge, the one against which the head of the T square is resting in Fig. 1, perfectly straight. Care should be taken to use the same side of the board each time, and to have the working edge always at the left.

T square—The T square consists of two parts, the head and the blade. The head of the square is shown in contact with the left edge of the board in

Fig. 1, with the blade extending across the paper. The blade is used as the ruling edge.

30 °-60 ° triangle—The angles formed by the edges of this instrument are 30°, 60°, and 90° as shown in Fig. 2; hence the name, 30°–60° triangle. Usually only one of the angles is used in referring to this triangle and it is called either a 30° triangle or a 60° triangle.

Fig. 2

45° triangle—This triangle is isosceles as shown in Fig. 3, having the equal angles 45° and the third angle 90°.

Scale—This instrument has U. S. standard graduations and is used for laying off dimensions. It should never be employed as a ruling edge.

Scroll—The scroll is made in several varieties and is used as a ruling edge for curves where the compasses cannot be employed.

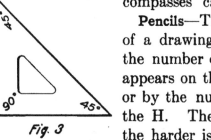

Fig. 3

Pencils—The degree of hardness of a drawing pencil is shown by the number of times the letter H appears on the wooden covering, or by the numeral that precedes the H. The larger the number, the harder is the pencil.

Erasers—That style in which the pencil eraser is at one end and the ink eraser at the other is best, simply because it reduces the number of articles on the drawing table.

Pencil-sharpener—Either a pencil file or block of sandpaper is necessary for sharpening the lead of the pencil. The former is preferable.

Thumb-tacks—Thumb-tacks with small heads should be selected, as they interfere less with the movement of the T square.

Drawing ink—An opaque, waterproof ink, one made especially for the purpose, should be used. Writing fluids are not suitable.

Penholder—The diameter of the penholder should be small in order that it may enter the small neck of the bottle in which drawing ink is usually furnished.

Writing pens—The ball point style of pen is best for lettering, as it is possible to get lines of approximately the same width whether making horizontal or vertical strokes.

The set of drafting instruments—While expensive instruments are not necessary, instruments with which accurate work may be done are essential. Should the student decide to purchase a set, the selection should be intrusted to one having had experience in their use.

A set of instruments complete enough for any draftsman is shown in Fig. 4.

The compass (A) used for drawing circles and arcs either with pencil or ink, is shown with the pencil attachment (B) in place. By loosening the clamping screw (C) the pencil attachment may be removed and the pen attachment (D) inserted.

The dividers (E) are used for dividing lines into a given number of equal parts also for transferring measurements from one part of a drawing to another.

The lengthening bar, shown at G, has a projection

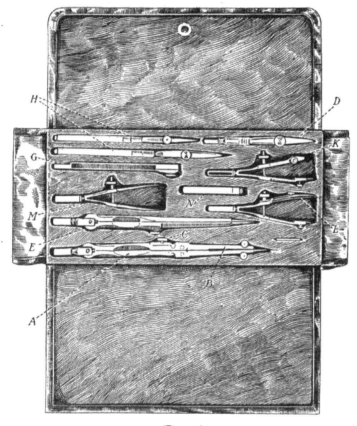

Fig. 4

at one end to be inserted in the leg of the compass, and a receptacle at the other end into which either the pencil or pen attachment may be placed. This extension is used when circles of large radii are desired.

Two sizes of straight-line pens are shown at H, either of which may be used, the selection being left to the workman.

Instruments K, L, and M are called bow instruments, K being the bow pencil, L the bow pen, and M the bow dividers. With these instruments small work may be executed more accurately than with the larger instruments.

N is a cylindrical tube used for holding leads for compasses.

CHAPTER II

PREPARATION AND USE OF MATERIAL

The pencils—To do good work, one must keep all pencils sharp; and to keep pencils sharp requires constant attention. The 6H pencil is to be sharpened at both ends; the 3H should be sharpened at one end only. To sharpen the pencil, remove the wood exposing about three-eighths of an inch of the lead, being careful, however, not to let the knife edge cut into the graphite. The lead should be sharpened by using the pencil sharpener or file. One end of the 6H pencil should be sharpened to an ordinary round point and the other end to a chisel point. To sharpen the round point after removing the wood, pass the lead across the file at the same time rotate the pencil between the fingers. This should give a long, conical point, tapering from the place where the wood meets the lead to the extreme point, with the end as sharp as a needle. When sharpening the chisel point, pass the lead across the file without rotating the pencil, tapering the cut from the wood to the end and removing the graphite until one-half at the end is gone. Repeat this process on the reverse side of the lead until the point has been brought to a knife edge. The 3H pencil should be sharpened to a conical point.

The 3H pencil is used for sketching, for printing, and for other free-hand work. All of the straight line mechanical work on the drawing should be executed with the 6H pencil. The round point of the 6H pencil is used for locating points and the chisel point for drawing lines, with the flat of the chisel against the ruling edge. One must early accustom himself to changing from the round to the chisel point, and vice versa, for different character of work.

The T square—The T square should always be used with the head against the left edge of the board, and the upper edge of the blade should always be employed as the ruling edge. All horizontal lines should be drawn from left to right, with the T square as a guide. The blade of the square should not be brought up to the point through which the line is to be drawn, but should be so placed that there will be a minute space between the blade and the line after the line shall have been drawn.

The triangles—For the vertical lines of the drawing the 60° triangle is usually employed, as it gives a longer ruling edge than the 45° triangle. When used for vertical lines, the triangle should be placed on the upper edge of the T square with the 60° angle at the right, as shown in Fig. 6. The placing of the triangle for 30°, 45°, and 60° lines will be determined by the direction in which these lines are to be drawn.

The scale—Keep the scale between the body and the line upon which the measurement is to be taken, thus bringing the instrument under the hand. When laying off dimensions, see that the mark is made directly opposite the required graduation on the scale.

The eraser—For pencil work use only the pencil end of the eraser.   If this will not do the work, it is no fault of the eraser but it is because the pencil lines have been made too heavy.   Erase in the direction in which the line is drawn and not across the line.   To remove a line, use many light strokes rather than dig into the paper with a few hard ones.   Never wet the eraser for either pencil or ink erasing.   For erasing inked lines use the ink end of the eraser, following the directions for erasing pencil lines.   Before attempting to erase, be sure that the ink is perfectly dry.

CHAPTER III

LAYING OUT THE SHEET

Each drawing should be inclosed within a rectangle, the lines of which may be called margin lines. This, in turn, should be inclosed in a larger one called the cut-off rectangle the size of the finished sheet. The cut-off lines are those upon which the drawing is trimmed after being taken from the board.

When laying out the margin and cut-off lines, proceed in the following manner:

1. Tack the paper to the board with the longest edge parallel with the upper edge of the blade of the T square. Place the paper nearer the upper than the lower edge of the board and nearer the left edge than the right. A good way to stretch the paper on the board is first to insert a thumb-tack in the upper left-hand corner, then, by passing the hand across the paper, being careful not to change the location of the paper on the board, stretch the upper edge and insert a tack in the upper right-hand corner. Stretching the left vertical edge, place a thumb-tack in the lower left-hand corner. The final stretching may be accomplished by passing the hand diagonally across the paper from the upper left to the lower right, and forcing the last tack into place. Remember that the tacks are thumb tacks and should not

be driven into the board by hammering with a knife
or T square.

2. Find the center of the sheet by using intersecting diagonals from the corners of the paper. See

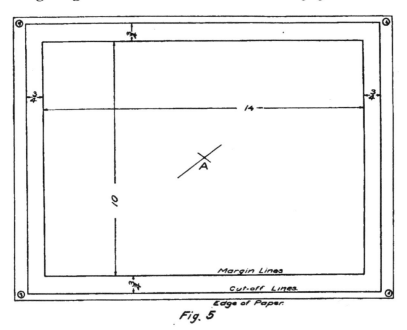

Fig. 5

point A, Fig. 5. Only that part of the diagonal at
or near the center of the sheet needs to be drawn.
The T square may be used in getting these lines,
by placing one of the edges of the blade diagonally so
it will intersect opposite corners of the paper.

3. Using the scale, measure five inches up and down
and seven inches to the right and left, locating points.
Most scales are not graduated to the extreme end,
and care should be taken to measure from the point

where the graduations begin and not from the end
of the scale. When locating the points, make a
very small dash directly opposite the required graduation on the scale. This mark should be made
with the round end of the 6H pencil, and should be
a very light dash and not a point drilled into the
paper. When using the scale, remember the directions, and keep the instrument under the hand.

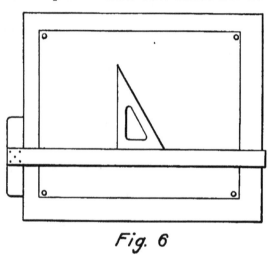

Fig. 6

4. With the head of the T square held firmly
against the left edge of the board, and using the upper
edge of the blade as guide, draw, very lightly, horizontal lines through the upper and lower points.
Through the points at left and right draw vertical
lines, using the 60° triangle as a ruling edge. When
using the triangle for vertical lines, keep the 60°
angle to the right, as shown in Fig. 6. Unless the
triangle is a large one, it will be necessary to make

the line in two parts. Use great care to make the line continuous, without a perceptible joint.

5. Locate points $\frac{3}{4}''$ above the upper margin line, $\frac{3}{4}''$ to the right of the right-hand margin line, $\frac{3}{4}''$ below the lower margin line, and $\frac{3}{4}''$ to the left of the left-hand margin line.

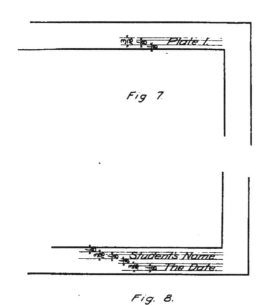

the line in two parts. Use great care to make the

Fig 7

Student's Name
The Date

Fig. 8.

6. Through these points draw lines forming the outer rectangle shown in Fig. 5. These lines form the finished size of the sheet.

In Figs. 7 and 8 are shown, respectively, the upper right and lower right-hand corners of the drawing, giving the lines to be drawn for the printing and the relation of these lines to the margin lines. The plates should be numbered consecutively, and each one should have the draftsman's name and the date of finishing the drawing.

CHAPTER IV.

USE OF INSTRUMENTS

(Straight Line Work)

PLATE 1

SUGGESTIONS FOR PENCILING

Accuracy and neatness are the two essentials for good work in mechanical drawing. As stated before, sharp pencils are absolutely necessary to do accurate work and should be the first things to receive attention when beginning the lesson. Not that only once during the lesson should the pencil be pointed, but one should begin with the tools in proper condition, and then,—keep them in condition. To do neat work requires clean hands, careful attention to details, light lines, and painstaking erasing, when erasures are required. Special attention should be given to the making of light, fine lines, for heavy lines are a disgrace to any draftsman. Remember that you are making the drawing on the paper and not into it. With the hard 6H pencil it requires no great pressure to cut a line into the paper that no amount of erasing will remove. One does not usually err by making the lines too fine and light. When using the scale, be careful to make the marks directly opposite the required graduations and when drawing the lines, see that they go exactly through the points located. All horizontal lines should be drawn from left to right and all vertical lines from the bottom up. In general, draw the lines away from the body.

The straight-line or drawing-pen—This pen should be used for inking all straight lines of the drawing and all curves where the scroll is required for the ruling edge. The thumb-screw is for adjusting the nibs of the pen for the different widths of lines. Care should be used not to screw this thumb-screw up too tight, for the threads are very fine and will easily strip off, thus ruining the pen. To fill the pen, place the quill—one is usually furnished with the small bottles of drawing ink—between the blades of the pen near the point, and the ink will readily flow from the point of the quill to the space between the nibs of the pen. Do not have more than $\frac{3}{16}''$ of ink in the pen, for a larger amount will cause a pressure at the point with a tendency to blot. Never hold the pen or quill over the paper, when filling the pen. Before attempting to make a line, see that no ink is on the outside of the blades of the pen. If there is, clean it off with the pen-wiper. If none is furnished with the bottle of ink, a soft cotton cloth may be used. Always try the pen outside of the cut-off line before starting to ink a drawing, not only to see if the pen is working properly, but also to adjust for the proper width of line. If interrupted while inking a drawing, always see that the pen is adjusted to the proper size of line, before taking up the work again. When it becomes necessary to re-fill the pen, the sediment remaining between the nibs should be wiped out by passing the pen-wiper between the blades. This means that the pen should be cleaned each time it is filled.

Using the drawing-pen—Hold the pen with the thumb-screw away from the body, with the end of the index finger of the right hand bearing against the outside blade just above the thumb-screw. Incline the pen slightly with the ruling edge and also in the direction of motion. The proper relation of pen and ruling edge is shown in Fig. 9. The distance between the pen point and the ruling edge should not be too great lest the outer nib of the pen be raised from the paper and make a ragged, uneven line.

Fig. 9

Erasing mistakes—If a mistake or blot makes it necessary to remove some of the inked work, first be sure that the ink is perfectly dry. Do not try to erase with a few hard strokes, for time and patience are necessary essentials to remove lines without showing the effect of the erasure.

PLATE 1

Follow the directions given on the next pages. Work carefully, accurately, and neatly. Keep the pencil sharp, using the conical point for locating points and the chisel point for drawing lines. Use a light, fine line.

INKING

When inking, follow this order:

1. Horizontal lines. Ink those at the top first.

2. Vertical lines. Ink those at the left first.

3. Oblique lines. Ink in the most convenient order.

4. Printing. Use the writing pen.

DIRECTIONS FOR MAKING PLATE 1

Locate a point $1\frac{9}{16}''$ below the upper margin line. Through this point draw a horizontal line connecting the left and right-hand margin lines. Locate points on this line $1\frac{11}{16}''$ in from each vertical margin line. Erase the portion of the line between these points and the margin lines. This should leave a horizontal line $10\frac{5}{8}''$ long. Draw two vertical lines extending down from the ends of this horizontal line a distance of $1\frac{1}{4}''$. Locate points $\frac{5}{16}''$ apart on the left-hand vertical line. Through the first point below the horizontal line, draw a dotted horizontal line to the vertical line at the right of the figure. Through the next point below, draw a full horizontal line. Through the next point below, draw a dotted line, and through the lowest point draw a full line, completing the figure. When drawing the dotted lines, make the dashes of equal length and have them equally spaced. Dashes should be not more than $\frac{1}{8}''$ nor less than $\frac{3}{32}''$ long, and the space between dashes should equal about $\frac{1}{3}$ the length of the dash.

Construct a square having sides $4''$ long, with the left side $2''$ from the left margin line and the lower side $1\frac{9}{16}''$ from the lower margin line. Using the scale, divide the lower side of the square into four equal parts. Through these points draw vertical lines to the upper side of the square. From the points where these vertical lines meet the upper side, draw 45° lines to the left side of the square. From the points on the lower side of the square, draw 45° lines to the left side.

Construct a square having sides 4″ long, with the right side 2″ from the right margin line and the lower side $1\frac{9}{16}$″ from the lower margin line. Divide the upper side of this square into four equal parts. Draw vertical lines across the square through the points located. Through the points on the upper side of the square draw 45° lines to the right side. Through the points where the vertical lines touch the lower side, draw 45° lines to the right side.

EXTRA PLATE

Draw a $7\frac{1}{2}''$ square in the center of the sheet, and copy one of the figures given on pages 22 and 23.

Note that the figures are determined by first drawing horizontal and vertical lines across the square from points equally spaced on the sides.

INKING

The light lines of the figure are to be drawn in pencil only; heavy lines are to be inked. When inking, follow the order given for Plate 1.

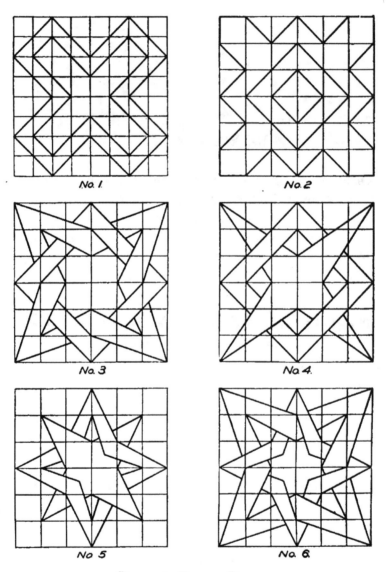

No. 1

No. 2

No. 3

No. 4.

No 5

No. 6

PLATE 1, EXTRA PLATE

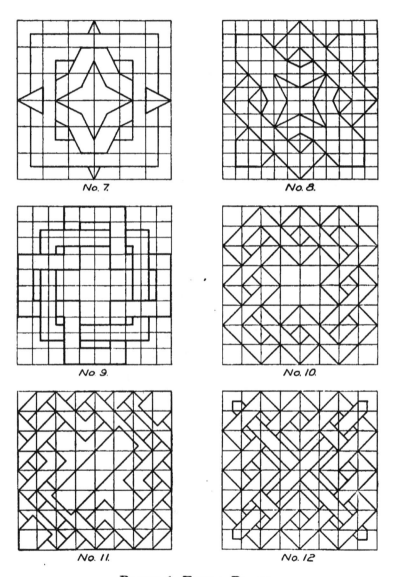

PLATE 1, EXTRA PLATE

CHAPTER V

USE OF TRIANGLES

PLATE 2

On pages 26 and 27 are shown the different ways
in which triangles may be placed in relation to the
T square, both singly and in combination. From
the illustrations it will be seen that angles of 30°,
45°, and 60° with the horizontal or vertical may be
obtained by using triangles singly, while for angles
of 15° and 75° a combination of triangles is necessary.
When two points are so located that the line connecting the two cannot be drawn with the triangles
in any of the positions shown on pages 26 and 27,
the quickest and best method is to place the pencil
on one of the points with one edge of a triangle
against the pencil, then, rotating the triangle until
the same edge coincides with the other point, draw
the line.

PARALLELS AND PERPENDICULARS

For drawing a parallel to a given line through a
given point, using the triangles, place one edge of a
triangle on the line with the second triangle bearing
against one of the other edges of the first triangle.
Holding the second triangle firmly, slide the first

one along its edge until it is in the required position. In Fig. 10 is illustrated the method whereby a line may be drawn through C parallel to M N. The long edge of the 45° triangle is made to coincide with the given line M N, and the long edge of the 60° triangle is placed against one of the short edges of the 45° triangle. These positions are shown in full lines. From this position the 45° triangle is moved along the edge of the 60° triangle until its long edge coincides with the point C, or to the position shown in dotted lines. The triangle is then in position to draw the line X Y, which will be parallel to M N.

To draw a perpendicular to a given line through a given point on or outside of the line, place the triangles in the same position as for parallels and then, rotating the 45° triangle to the position shown in dotted lines in Fig. 11, the required perpendicular may be drawn. In Fig. 11, the line M N is the given line, and X Y the perpendicular through either C or C'.

Another method which may be employed for drawing the perpendicular is illustrated in Fig. 12. In this case the short edge of the triangle is made to coincide with the given line and is then pushed along the edge of the second triangle, thereby bringing the other short edge to the point through which the required perpendicular is to be drawn. In Fig. 12, the line M N is the given line and X Y is the required perpendicular through C.

In Figs. 10, 11, and 12 the 45° triangle was used as the first triangle, although the 60° triangle might have been used with the same result. Thus in Fig. 13

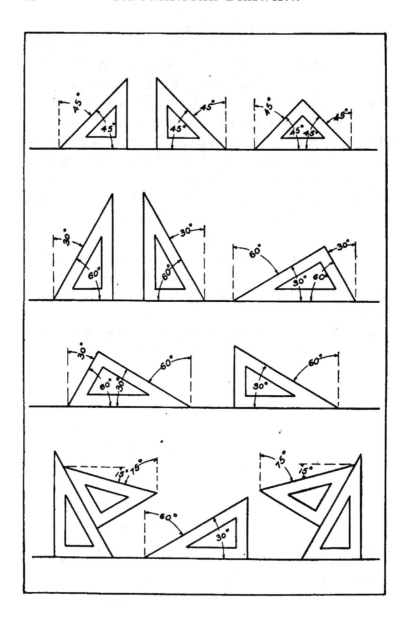

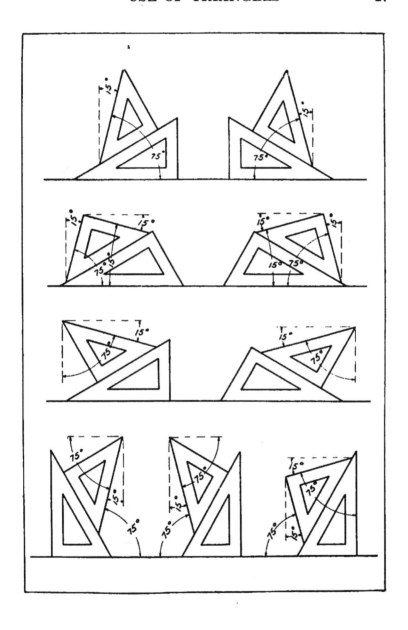

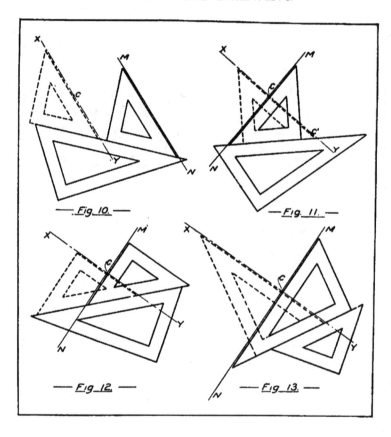

we have the same proposition as in Fig. 11, with an interchange of triangles.   Do not under any consideration place one edge of the right angle of a triangle against a line and use the other edge bounding the right angle to draw a perpendicular to that line. Good, accurate work cannot be done in this way.

PLATE 2

Make drawings of the geometrical figures given in one of the rectangles shown on page 30. The rectangle shown in the copy represents the margin lines.

Keep the pencil sharp, using the conical point for locating points and the chisel point for drawing lines. Make very light, fine lines. Do not put the dimensions on the drawing.

INKING

Ink only the heavy lines shown in the copy. Have the drawing complete in pencil before inking.

When inking, follow the order given on page 18.

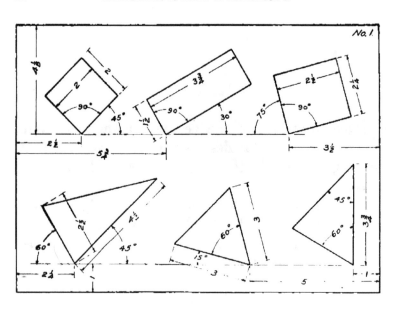

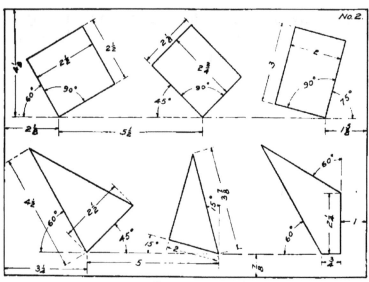

PLATE 2

CHAPTER VI

WORKING DRAWINGS

PLATES 3 AND 4

If we look at an object through a transparent plane and trace the outline as seen upon that plane, the result is a perspective drawing of the object. The relation of the object to the plane determines the character of the drawing, and as the observer changes his position relative to the plane his view of the object will also change, giving for different positions entirely different drawings.

In perspective drawing all the points of sight producing the outline on the picture plane converge to one point, namely the eye of the observer. This causes all lines of the object in contact with the picture plane to be in their true length in the drawing, while all lines back of the picture plane will be shortened. This foreshortening of lines with the inability to measure them, makes the perspective drawing of very little value to the workman.

In making the working drawing or mechanical drawing, we consider that all lines of sight producing the picture are parallel to each other, thus giving views of lines parallel to the picture plane in their true length. In Figs. 14 and 15, page 32, are il-

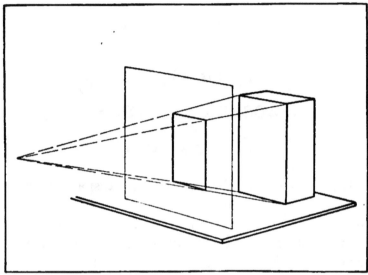

Fig. 14

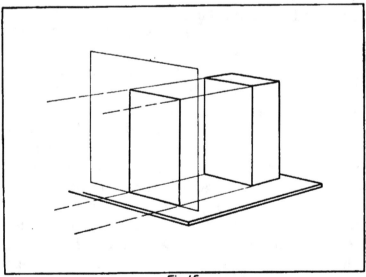

Fig. 15

lustrated the two principles by which the perspective drawing and the working drawing views are obtained. Note that in Fig. 14 none of the lines are seen on the picture plane in their true length, being foreshortened because of the converging lines of sight, while in Fig. 15 all the lines of one face of the object are seen not only in their true length but also in the true relation to each other.

The perspective drawing gives the general outline and relation of parts in one view, while in mechanical drawing more than one view is required to clearly illustrate an object. Sometimes two views only are necessary to show the shape or construction, while in other cases three, and occasionally more, are required. These two or more views, drawn according to given principles and in the proper relation to each other, with enough dimensions for making the object represented, constitute a mechanical drawing.

THE THREE VIEWS

A free-hand perspective drawing of a model from which it is required to make the working drawing, is given in Fig. 16, page 34. Let us, for convenience, call the surface A the front surface of the object, B the top surface, and C the side surface. To get a view of surface A, we would look in the direction indicated by arrows D, shown in Fig. 17. This view, being of the front of the object, may be called the front view, and may be placed at A, Fig. 18. The view of the top surface, or top view, would be ob-

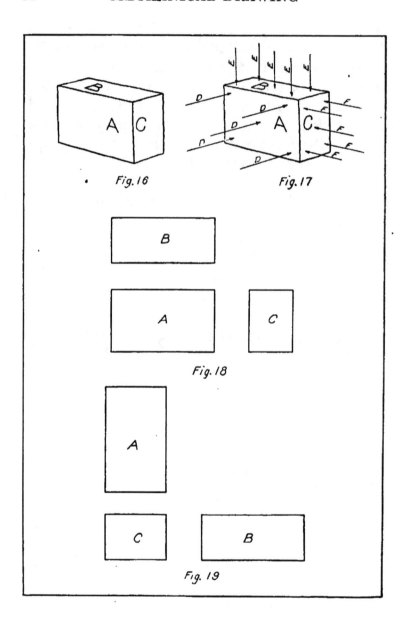

Fig. 16

Fig. 17

Fig. 18

Fig. 19

tained by looking in the direction indicated by the arrows E. This being the top view, we would most naturally place it above the front view, when grouping the views. In Fig. 18, the top view is shown at B, directly above the front view. To obtain the view of the side of the object, or surface C, we would look in the direction indicated by the arrows F, and would get the view shown at C, Fig. 18. This being the right side view of the object, it is placed at the right of the front view. The relation of the three views then will be as follows:

The top view is directly over the front view.

The side view is directly to the right of the front view.

A study of the three views given in Fig. 18 shows that the height of the front and side views is the same, the breadth of the front and top views is equal, and that the height of the top view and breadth of the side view are identical. From these statements we may formulate the following rules:

1. *The vertical dimensions on front and side views are equal.*

2. *The horizontal dimensions on front and top views are equal.*

3. *The vertical dimension on the top view equals the horizontal dimension on the side view.*

It is not necessary to use the front face of the object as the front view. Any surface may be employed as the front view, provided the other views are drawn in the proper relation to this view. Thus, in Fig. 19, the surface C, drawn with its long edge in a horizontal position, is used as the front view, with surfaces

A and B as top and side views, respectively. Still another combination is shown in Fig. 20, page 37, in which B is the front view, C the side view, and the surface across the object from A, which may be called A′, the top view. In Fig. 21, the front view C in Fig. 19 is drawn with its short edge horizontal. This causes B to become the top view and A the side view. Fig. 22 is the same as Fig. 20, with the top view omitted. Notice that these two views show the shape and size of the object as well as do the three views of Fig. 20. This saves the drawing of the third view; but in our practice, for the present, we will draw the three views even though they may not be absolutely necessary.

The block shown in Fig. 23, page 38 is the same as that illustrated in Fig. 16, but with a mortise cut into the block on the surface A. Three views of this object are given in Fig. 24, in which the surface A becomes the front view, B the top view, and C the side view. The mortise is represented on the front view by the inner rectangle, and is expressed on the top and side views by dotted lines. Inasmuch as the mortise is not visible from the top or side, some characteristic must be employed to distinguish the visible from the invisible edge. Full lines are employed to represent visible edges of the object, and dotted lines for the invisible edges. These lines may be grouped under one head and called the main or primary lines of the drawing.

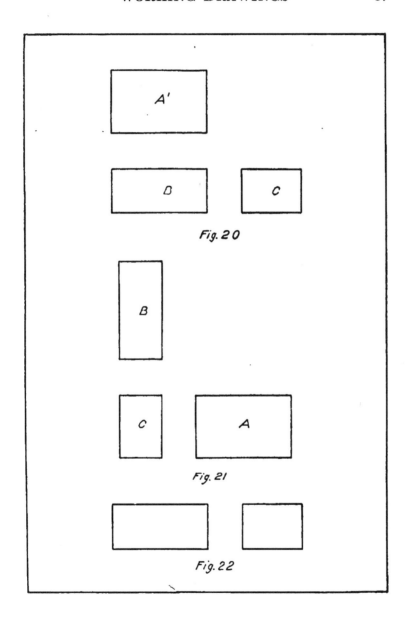

Fig. 20

Fig. 21

Fig. 22

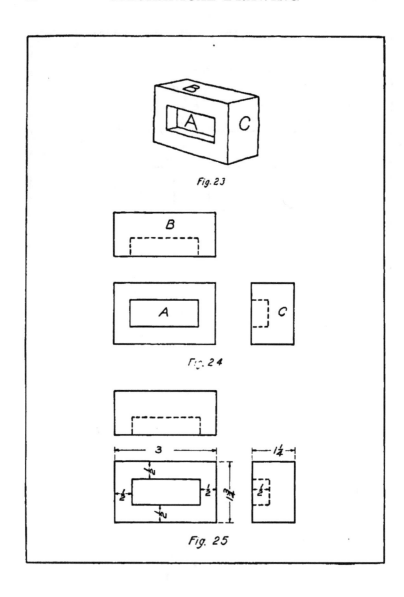

Fig. 23

Fig. 24

Fig. 25

DIMENSIONS

The drawings we have just considered, while well representing the objects, do not give enough information for making the models. Not only must the general outline, shape, etc., be shown, but the dimensions necessary for making the model to a definite size must also be given. In Fig. 25 is shown the complete drawing of the model illustrated in Fig. 24, not only the shape but the sizes as well being given. To indicate these dimensions, other lines and characters are employed which may be grouped under one head and called the secondary lines of the drawing. The names of the lines included in this secondary group are *witness or extension lines,* and *dimension lines.* · Other characters used are arrow-heads. The figures placed on the drawing are called dimensions.

A dimension line is one upon which the dimension is placed and in which a break is made for the dimension. Arrow-heads are placed at the extremities of this line. A witness or extension line is one extending out from the object line to and a little beyond the dimension line, employed when the dimension is placed outside the view. Different characters of lines are used in different drafting rooms for these secondary lines, so a system employed in one text book could not conform to every drafting-room system. For the work in this course, dotted lines will be used for witness lines, and a full line, with space reserved for the dimension, will be employed for the dimension line. The arrow-head should not fall short of or project over the witness or object line to which it goes, and should be made *short, narrow, and pointed.*

SUGGESTIONS FOR DIMENSIONING

Place all dimensions possible on one view, but give dimensions to full lines in preference to dotted.

Avoid crossing horizontal and vertical dimension lines.

So place the dimension that it can be erased without erasing a line of the drawing.

Have horizontal dimensions read from the bottom and vertical dimensions from the right of the drawing. See Fig. 25.

Do not use an oblique line as the vinculum of a fraction; make this line parallel to the dimension line. Make $\frac{1}{8}''$, not 1/8''. Use great care in making the figures; these are vital parts of the drawing.

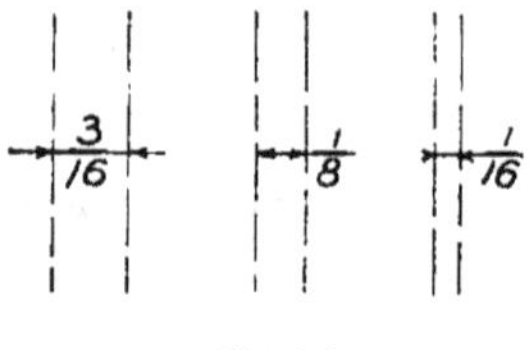

Fig. 26

Be sure to have the three dimensions—length, breadth, and thickness—for the main piece, also for all projections and recesses.

When it is impossible, because of the narrowness of the space, to get both dimensions and arrow-heads between the witness lines, the arrow-heads, or the dimension, or both, may be placed outside. These three methods are shown in Fig. 26.

Do not use the '' marks, when all dimensions are in inches.

LOCATING THE DRAWING IN THE RECTANGLE

Unless a drawing is located in the center of the rectangle in which it is drawn, the appearance of the drawing is marred. It is, therefore, advisable to place the views with reasonable spaces between, and with equal margins at the top and bottom, also at the

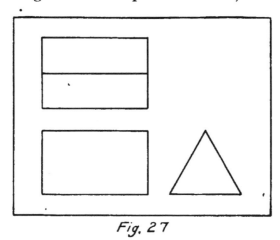

Fig. 27

right and left. The drawing shown in **Fig. 27**, surely is more pleasing to the eye and makes a better impression than the one shown in Fig. 28. This is due entirely to the proper placing of the drawing within the rectangle bounding it.

The exact location of the drawing on the sheet should be determined before a line is drawn. This may be obtained by finding the full height and width of the drawing, including the space between views, subtracting these dimensions from the height and width of the sheet, and dividing the difference by

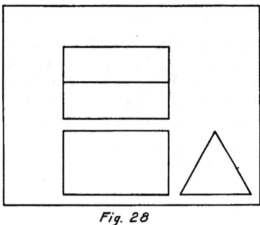

Fig. 28

two. This will give the spaces at the top and bottom and the spaces at the right and left. Thus, in

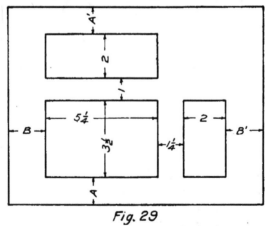

Fig. 29

Fig. 29, the spaces A and A′ may be obtained by adding the height of the front view, $3\frac{1}{2}''$, the height of the top view, 2″, and the space to be allowed between the front and the top views, 1″, and subtracting the sum, 6½″, from the height of the rectangle, 9″, leaving the amount to be divided equally between A and A′, 2½″. If we make A and A′ equal, then each would be 1¼″. The spaces B and B′ may be obtained in the same manner by adding the width of the front view, 5¼″, the width of the side view, 2″, and the space between views, 1¼″, making a total of 8½″. Subtracting this total from the width of the rectangle, 12″, and dividing by two, we have, $12″ - 8½″ = 3½″$, and $3½″ \div 2 = 1¾″$. Therefore

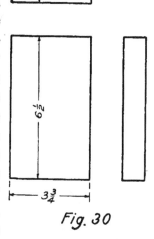

Fig. 30

spaces B and B′ should each be made 1¾″. It will be noticed in the above illustration that more space was allowed between the front and side views than between the front and top views. It is not necessary that these spaces be equal; in fact it is better under certain conditions that they be unequal. In the figure we have been discussing, there is more blank space horizontally than vertically; therefore there should be more space between views horizontally than vertically.

Let us consider one other condition. The drawing shown in Fig. 30 is to be placed in a rectangle 9″ high and 12″ broad. If we should allow 1″ between the front and side views, the total horizontal width of the drawing would be 5¾″. This sub-

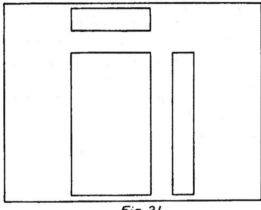

Fig. 31

tracted from the horizontal dimension of the sheet would give 6¼″ to be divided into two equal parts, for the spaces at right and left. These spaces would then be 3⅛″ each. Allowing 1″ between the front and top views, the total height of the drawing would be 8½″. The difference between this height and the height of the rectangle would be ½″, which, divided by two, would give ¼″ each for spaces at the top and the bottom of the drawing. A drawing placed according to these conditions is shown in Fig. 31. One can readily see that there is too much space between the top and front views, when we compare it with the spaces at the top and the bottom of the drawing. If we reduce this space from 1″ to ½″, we will then have an additional ½″ to be divided equally and added to the spaces at the top and the bottom, making each of these ½″, instead of ¼″ as

in the previous case. A drawing made to suit these
changed dimensions is shown in Fig. 32. Even

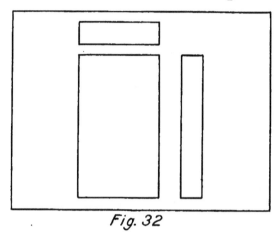

Fig. 32

though this is an improvement over the other condition, it still looks awkward with so much space

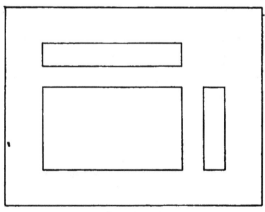

Fig. 33

horizontally and so little vertically. The reason
for this is that we have been attempting to place the

long dimension of the drawing in the short dimension of the rectangle. By a rearrangement of views as shown in Fig. 33, thereby placing the long dimension of the drawing parallel to the long dimension of the rectangle, a much more satisfactory result is secured. It will, then, in all drawings, be necessary to find which is to be the longest dimension, and place this the long way of the sheet.

INKING THE WORKING DRAWING

It is well to divide the lines of a drawing into groups before starting the inking, that systematic and time saving work may be done. The primary lines of the drawing—object lines—may be placed in one group, the secondary lines in another group, and the free-hand work, which would include dimensions, arrow-heads, and printing, in a third group. To distinguish the primary from the secondary lines, different colored inks are often employed. Black is always used for object lines, and red is often employed for witness and dimension lines. Many times the black is used for both. In the latter case the distinguishing element is the character and size of the line.

When inking drawings similar to those we have been studying, observe the following order:

ORDER FOR INKING

Group 1. Object lines; heavy lines, with black ink.
Horizontal lines; the upper ones first.
Vertical lines; those at the left first.
Oblique lines; the most convenient way.

Group 2. Witness and dimension lines; light lines,
with red ink.
Horizontal lines; the upper ones first.
Vertical lines; those at the left first.
Oblique lines; the most convenient way.

Group 3. Arrow-heads, dimensions and printing;
with black ink.
Free-hand, with writing pen. Work
from the upper left hand corner to
the lower right.

Group 4. Margin lines; heavy lines, with black ink.
When these are made the same size as
the object lines, they should be inked
with Group 1.

PLATE 3

After laying off the margin and cut-off lines, divide the sheet into two equal rectangles by a vertical line through the center. Two problems are to be solved on this sheet, one at the left of the vertical line, and one at the right, to be taken, respectively, from the figures shown on pages 49 and 50. The problems will be selected by the instructor.

Make three views, complete with dimension lines and dimensions. If possible, leave at least 1″ between the views, and locate the drawing in the rectangle with equal spaces at the right and the left, also at the top and the bottom. Use light pencil lines.

INKING

Have the drawing complete in pencil before inking. When inking, follow the order given on page 47.

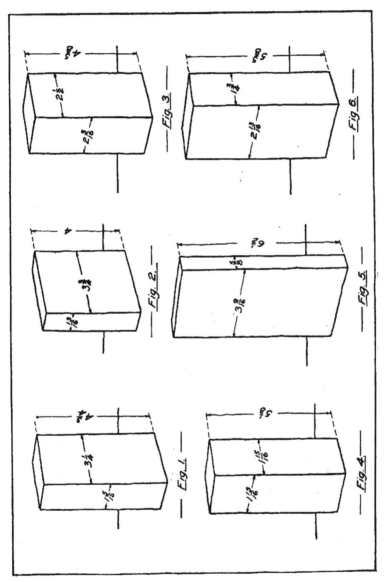

PLATE 3

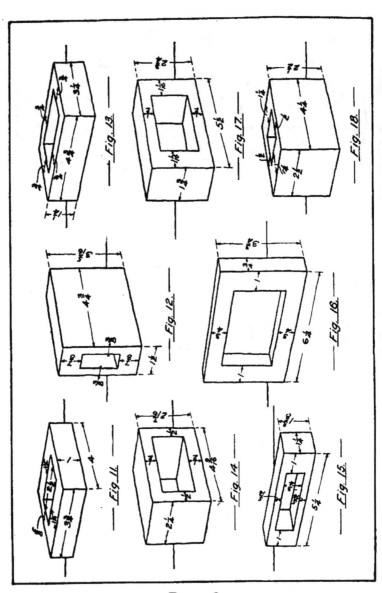

PLATE 3

EXTRA PLATE

Divide the rectangle made by the margin lines into two equal parts by a vertical line through the center. The problem to be drawn in the rectangle at the left will be selected from the upper part of page 52, that for the right-hand rectangle from the lower part of the same page.

Make three views, complete with dimension lines and dimensions. If possible, leave at least 1″ between the views and locate the drawing in the rectangle with equal spaces at the right and the left, also at the top and the bottom. Use light pencil lines.

INKING

Have the drawing complete in pencil before inking. When inking, follow the order given on page 47.

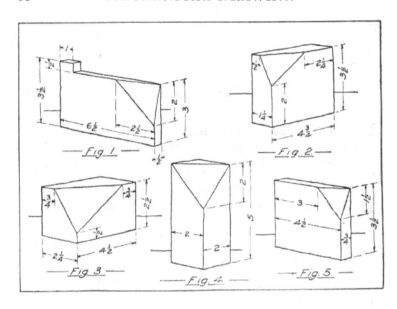

Fig. 1. Fig. 2. Fig. 3 Fig. 4 Fig. 5.

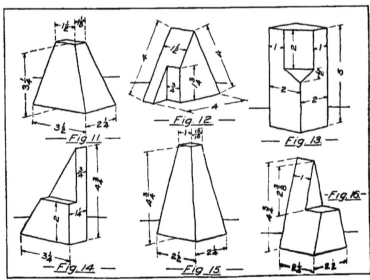

Fig. 11 Fig. 12. Fig. 13. Fig. 14 Fig. 15 Fig. 16.

PLATE 4. EXTRA PLATE

PLATE 5 *

Make three views—top, front, and right side—of one of the pieces shown in perspective on pages 62 to 64. The problem will occupy the entire sheet. Leave from 1″ to 1½″ between the views, and locate the drawing on the sheet with equal spaces at the right and the left, also at the top and the bottom.

INKING

Have the drawing complete in pencil before inking. When inking, follow the order given on page 47.

*NOTE. Plate 6 is to be finished before Plate 5 is taken from the board.

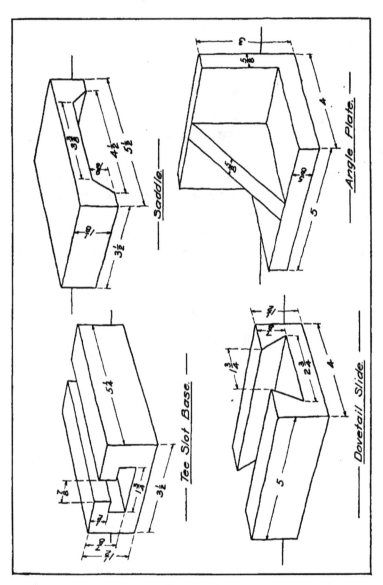

PLATE 5

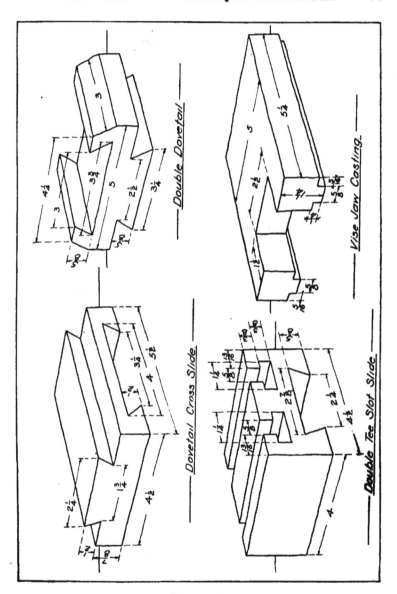

PLATE 5

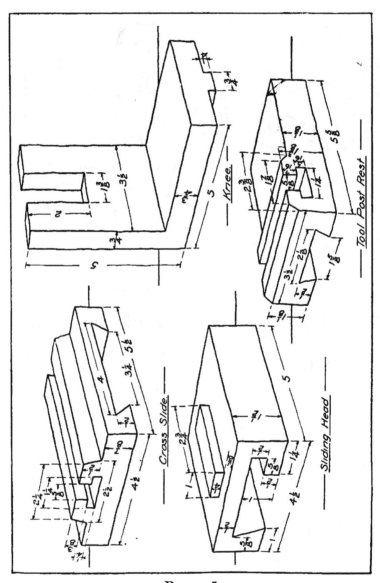

PLATE 5

PLATE 6

Make a tracing from Plate 5. When inking, follow the order for inking given on page 47, except the directions for the witness and dimension lines. On the tracing make the lines included in Group 2 black and very fine. This will require that the primary and secondary lines be inked in separate groups, as when the colored ink is used; for the former will be heavy lines and the latter fine lines. As the width of the line is the only distinguishing characteristic, it will be necessary to have the difference in width quite marked.

CHAPTER VIII

ASSEMBLY DRAWINGS

PLATE 7

All of the drawings which we have made up to the present time have been of a single piece and are called detail drawings. When two or more pieces, made to fit together, are drawn as they would be when they are put together, we have an assembly or construction drawing. The principles involved in working out the views are the same as for the single piece, although more care is required in the selection of views, in order to show the construction clearly and at the same time to use as few dotted lines as possible. Dotted lines on a drawing are always confusing, and in many cases one combination of views will mean fewer dotted lines than another. A little time and study will determine the best combination.

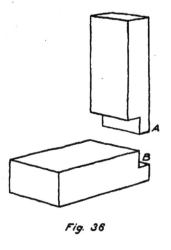

Fig. 36

In Fig. 36 are shown two parts of a joint used in woodworking, in which the tenon, A, of one piece fits into the recess cut into the other piece at B. An assembly drawing of this same joint, using three views, is shown in Fig. 37. In reality, two views will show the details of construction clearly. These two

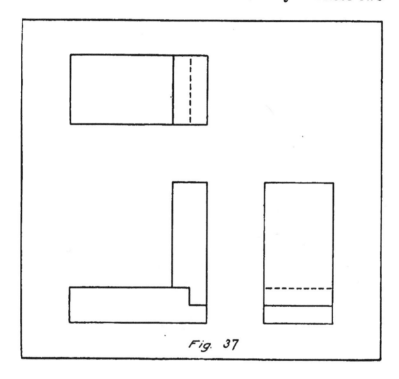

Fig. 37

views, with the dimensions for making each part, are illustrated in Fig. 38. When dimensioning a drawing of this character, dimension each of the component parts independent of the other. In Fig. 38, the dimensions A, B, C, D, and E are complete for making the horizontal piece and should be put on before any attempt is made to dimension the vertical

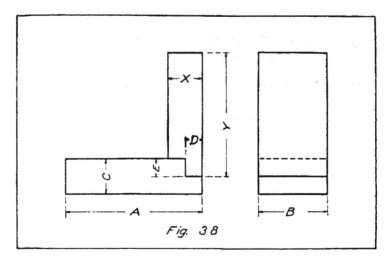

Fig. 38

piece. When placing the dimensions on the vertical
piece, it will be found that some of the dimensions
used for the horizontal piece may also be used for
the vertical. Thus the dimension D will serve not
only as the width of the cut on the horizontal piece,
but also for the tongue on the vertical piece; and the
dimension B will serve equally well for the width of
either piece. This leaves simply the dimensions X
and Y to be added, in order to complete the dimensioning of the vertical piece, although the full dimensions would include X, Y, B, D, and E.

PLATE 7

Divide the rectangle made by the margin lines into two rectangles by a vertical line through the center. In the rectangles on pages 70 to 73 are shown perspective drawings of parts of different woodworking joints. Make three views—top, front, and right side—of two of these problems. When making the drawings, show the constructions clearly, and choose the views showing the fewest dotted lines. Leave about 1″ between the views, and locate the drawing in the center of the rectangle.

INKING

Have the drawing complete in pencil before inking. When inking, follow the order given on page 47.

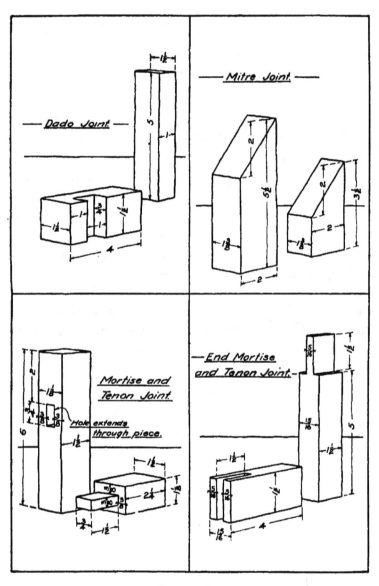

PLATE 7

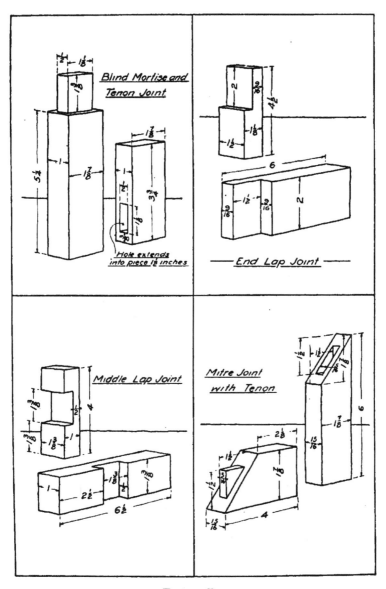

PLATE 7

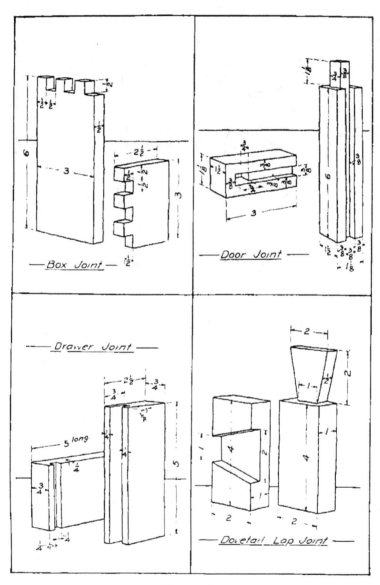

PLATE 7

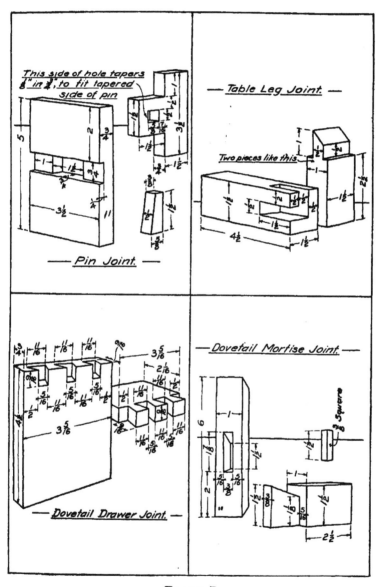

PLATE 7

EXTRA PLATE

On pages 75 and 76 are given detail drawings of parts of woodworking models. Make assembly drawing, showing the construction in full lines, if possible. One problem will occupy an entire sheet.

INKING

Have the drawing complete in pencil before inking. When inking, follow the order given on page 47.

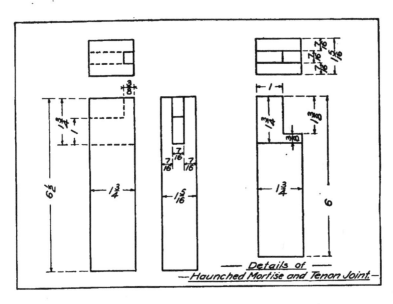

— Details of —
— Haunched Mortise and Tenon Joint. —

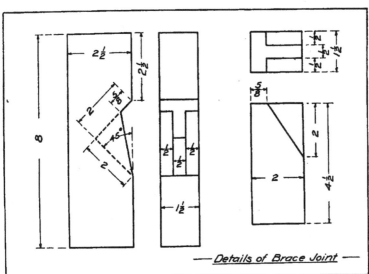

— Details of Brace Joint —

PLATE 7. EXTRA PLATE

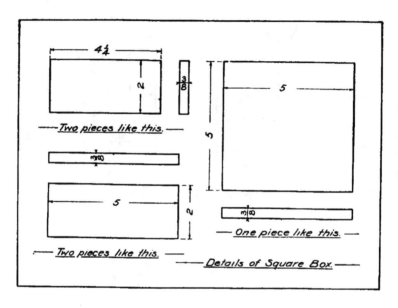

Details of Square Box.

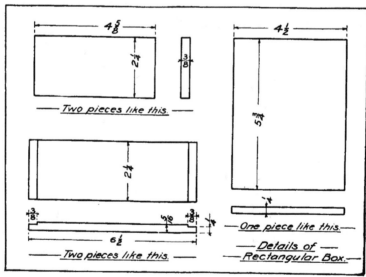

Details of Rectangular Box.

PLATE 7.   EXTRA PLATE

EXTRA PLATE

On pages 78 and 79 are given the assembly drawings of some woodworking joints.   Make the detail
drawings.

INKING

Have the drawing complete in pencil before inking.
When inking, follow the order given on page 47.

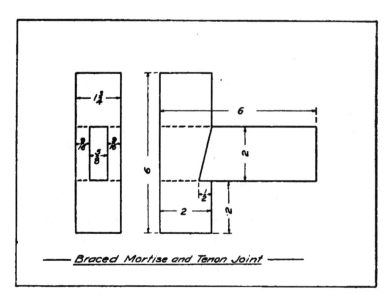

Braced Mortise and Tenon Joint

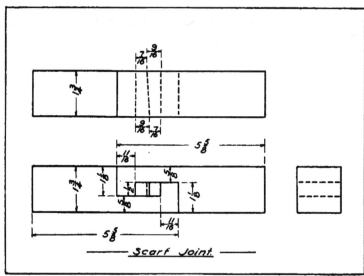

Scarf Joint.

PLATE 7. EXTRA PLATE

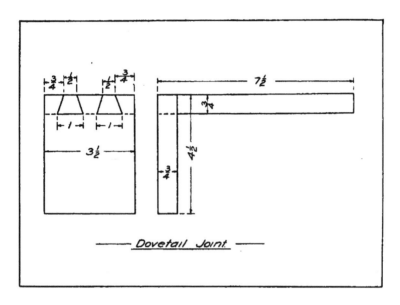

Dovetail Joint

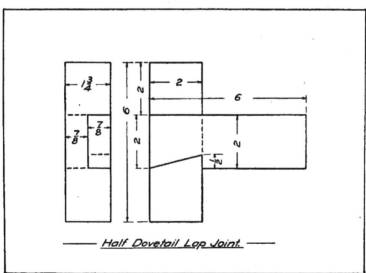

Half Dovetail Lap Joint.

PLATE 7. EXTRA PLATE

CHAPTER IX

USE OF INSTRUMENTS

(Curved Work)

PLATE 8

The compasses—As before stated, the compasses are provided with pencil and pen attachments and a lengthening bar. The lengthening bar is inserted between the fixed leg of the instrument and either the pencil or pen attachment, when it is desired to draw an arc whose radius is greater than the capacity of the regular instrument. For circles of large radii, the knuckle joints in the legs of the instrument should be bent to bring both needle point and pencil or pen attachment at right angles to the paper. This is absolutely necessary when the pen attachment is used, in order that both nibs of the pen come in contact with the paper.

The compasses should never be placed on the scale when setting for a radius; this will eventually make holes in the scale, thereby ruining its accuracy. Measure to the left from the center about which the circle is to be drawn a distance equal to the radius of the circle, and after locating the needle point exactly at the center set the pencil point to the mark through which the arc is to be drawn. Use only one hand when drawing a circle, grasping the instrument by the handle where the two legs are jointed. Rotate the compasses from the left, up over the top, down through the right and bottom to the starting point. Always rotate the compasses in this direction, beginning at the left. Go over the line only once, making a light, fine line. An attempt to make a heavy line will probably result in spreading the legs of the instrument, thus increasing the radius.

The dividers—While the dividers are used for transferring distances from one part of the drawing to another part, they should never be employed for transferring a measurement from the scale to the drawing. Measurements on the drawing should be made directly from the scale. The dividers are most useful for dividing a line into a given number of equal parts, when the scale cannot be employed for the purpose. When thus used they should be held by the handle at the top of the instrument and should be rotated first on one side of the line and then on the other. Should the first setting not divide the line equally, a second setting, and probably more, will be required. If the divider is provided with a hair-spring adjustment, this will be found very useful for adding to or subtracting from the original setting. Do not push the points of the instrument through the paper until the exact setting has been obtained.

The bow instruments—These instruments are used on small work for the same purposes as the large ones are employed for large work, the directions for their use being the same as those given for the large compasses and dividers. The bow instruments should be employed for everything within their capacity.

PLATE 8

Follow the directions given on the following pages. Use extreme care when working out the figures. Follow the directions for the use of the compasses and dividers.

INKING

The margin lines are the only straight lines to be inked. These should be inked last. All the arcs, with the exception of the circle about the point B, are to be inked. When inking the arcs, begin with the largest.

DIRECTIONS FOR MAKING PLATE 8

Three inches below the upper margin line, draw a light horizontal line connecting the left and right-hand margins. On this line and $2\frac{1}{2}''$ from the left-hand margin, locate a point. Letter this point A. Locate a point, to be lettered B, $4\frac{1}{2}''$ to the right of point A; and $4\frac{1}{2}''$ to the right of B locate a point, which we will call X. Through the points A, B, and X draw vertical lines about 5″ long, extending equidistant above and below the horizontal line.

Draw a horizontal line $2\frac{1}{2}''$ above the lower margin line, connecting the left and right hand margin lines. On this horizontal line and 2″ to the left of the right hand margin line, locate a point. This point we will call D. Beginning with point D and measuring to the left, locate four more points $2\frac{1}{2}''$ apart. Call these points E, F, G, and H. Through these points draw vertical lines about 1″ long, projecting equally above and below the horizontal line.

With point A, on the upper horizontal line, as a center, draw a circle having a 4″ diameter. Using the same center, draw a dotted circle of $1\frac{1}{2}''$ radius. When drawing the dotted line, make the dashes of equal length and have them equally spaced. Dashes should not be more than $\frac{1}{8}''$ long and the space between dashes should equal about $\frac{1}{3}$ the length of the dash.   Again using the center A, draw a circle of 1″ radius.   This circle is to be drawn a full line. Concentric with the three circles just drawn, make a dotted circle of 1″ diameter.

With a 1″ radius, draw a circle about the point B. Using the 45° triangle, draw lines through the center of this circle, dividing the circumference into eight equal parts.   Beginning at the point where the vertical line through the center cuts the upper part of the circle and passing around to the right, number these points from 1 to 8, consecutively.   With point 1 as a center, and with a radius equal to the distance from 1 to B, draw a semicircle to the right of the vertical line.   Using point 2 as a center, with the same radius, draw an arc from a point where it would intersect the semicircle about point 1, to the right, until it strikes point B.   Draw about point 3, to the right, with the same radius, an arc from the point of its intersection with the arc about point 2, until it reaches point B.   Complete the figure, using, in succession, points 4, 5, 6, 7, and 8.   After the arc about point 8 is drawn, it will be necessary to extend the arc drawn about point 1 in order to make this arc the same as the others.

With X as a center, draw a circle having a diameter of 4″.   Using the bow dividers, divide the horizontal diameter into six equal parts.   Beginning at the left, at the point where the horizontal line cuts the arc, letter the points a, b, c, d, e, f, and g.   Point d will coincide with X.   With b as a center, and a radius equal to the distance from b to a, draw a semicircle above the horizontal line. This semicircle should touch the point c. With the same radius, draw a semicircle below the horizontal line, using f as a center. ·This semicircle should touch the points e and g. With c as a center, and a radius equal to a c, draw a semicircle above the horizontal line. This semicircle should touch the point e. Using the same radius, draw a semicircle below the horizontal line, with the point e as a center. This semicircle will touch g and c.

With points D, F, and H, on the lower horizontal line as centers, with a radius of $1\frac{1}{2}''$, draw semicircles above the horizontal line. With E and G as centers, and the same radius, draw semicircles below the horizontal line. Again using points D, F, and H as centers, draw semicircles with a 1″ radius above the horizontal line. Below the horizontal line draw semicircles about points E and G, using a 1″ radius.

EXTRA PLATE

Draw a $7\frac{1}{2}''$ square in the center of the sheet, and work out one of the figures shown on page 87. The centers for the arcs are determined by drawing horizontal and vertical lines across the square from points equally spaced on the sides.

INKING

Only the heavy lines of the figures are to be inked. Ink the large arcs first.

No. 1.

No. 2.

No. 3.

No. 4.

No. 5.

No. 6.

PLATE 7. EXTRA PLATE

EXTRA PLATE

Draw one of the figures shown on page 89. All the lines radiating from the centers C may be obtained by using the triangles on the T square, either singly or in combination. Work carefully and accurately. Do not put the dimensions on the drawing.

INKING

Ink only the heavy lines of the figure. Ink all the arcs first, and then the straight lines.

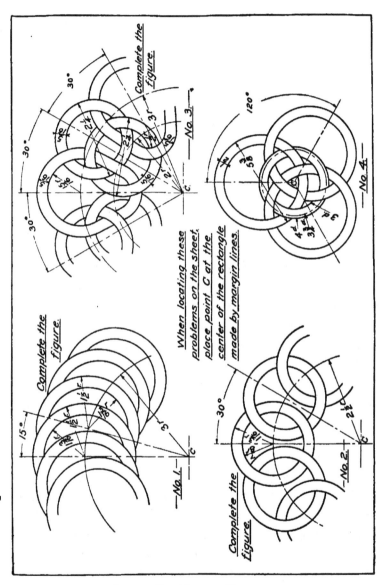

PLATE 7. EXTRA PLATE

CHAPTER X

CYLINDRICAL WORK

PLATES 9 AND 10

Whenever a circular piece or circular hole is shown in a mechanical drawing, a line is used to indicate the location of the center of the piece or hole. This line is known as a center line, and is included in the group of secondary lines already referred to. It represents the axis of revolution of the piece and becomes the axis of symmetry in the drawing. In Fig. 39 are shown two views of a double lever. The main center line of these views is the horizontal line marked "primary center line." This is the center line of the main cylinder about which the model is constructed, and also serves as a center line for the hole through this main cylinder. As the cylinder and hole are concentric, the one center line will answer for both.

On the circle view of a cylinder or a cylindrical hole, two center lines, both passing through the center of the circle, are drawn. These two lines are usually at right angles to each other. Thus, in Fig. 39, the vertical center line is drawn at right angles to the other main center line. This vertical center line is also a primary center line of the drawing, and, had the top view been drawn, would have extended up

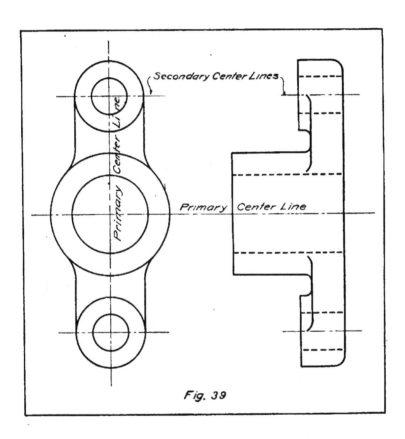

Secondary Center Lines

Primary Center Line

Primary Center Line

Fig. 39

through that view, connecting it with the front view, as the horizontal center line connects the front and side views.

The two small cylinders on the arms extending out from the main cylinder also require center lines. As these are not the main center lines about which the drawing is made, they do not connect the two views, although they have to be shown on both views. These may be called secondary center lines. The vertical center line serves not only as the main center line for that view, but also as one of the center lines of the small cylinders and the holes in these cylinders.

Alternate long and short dashes may be used to distinguish the center line from the other secondary lines of the drawing, although, as with all of these lines, this characteristic is dependent upon the systems used in the different drawing-rooms in which the drawings may be made. These lines should be inked with the other secondary lines, and should be the same width and color.

When a series of holes equidistant from a given point are to be represented, a line, called a circular center line, is employed. As its name implies, it is a circle drawn about the point from which the holes are equidistant, passing through the centers of the holes. The line X, Fig. 40, is a circular center line answering as one of the center lines of all the small holes represented on the front view. The other center line for these holes is a portion of a radial line drawn through the center of the circle representing the hole. These lines should not be drawn to the center of the piece, but should extend only through the circles for which they are the center lines. This is shown clearly in Fig. 40. Notice also on this figure the center lines for the small holes on the side view.

The primary center lines should be the first pencil lines to be drawn, and should be so located that when

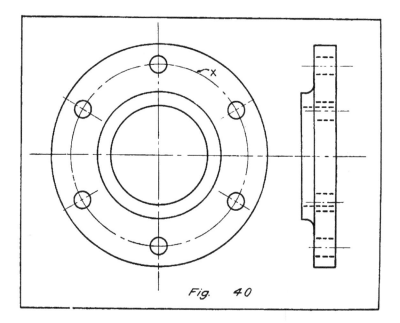

Fig. 40

the drawing is complete it will be in the center of the rectangle. After drawing the center lines, the measurements should be made from these and the drawing built up about them.

The introduction of the curved lines of the object and the center lines necessitates a new order for inking the drawing. When inking drawings similar to those shown in Figs. 39 and 40, follow the order given below:

ORDER FOR INKING

Group 1. Object lines; heavy lines, with black ink.
Arcs; begin with the largest.
Horizontal lines; the upper ones first.
Vertical lines; those at the left first.
Oblique lines; the most convenient way.

Group 2. Witness, center, and dimension lines; light lines, with red ink.
Arcs; begin with the largest.
Horizontal lines; the upper ones first.
Vertical lines; those at the left first.
Oblique lines; the most convenient way.

Group 3. Arrow-heads, dimensions, and printing; with black ink.
Free hand, with writing pen. Work from the upper left to the lower right.

Group 4. Margin lines; heavy lines, with black ink.
When these are made the same size as the object lines, they should be inked with Group 1.

PLATE 9

Divide the rectangle made by the margin lines into two equal rectangles by a vertical line through the center. In the left-hand rectangle draw two views of one of the objects shown in perspective on page 96, and in the right-hand rectangle draw two views of one of the models shown on page 97.

Select the views having the fewest dotted lines, consistent with a clear representation. Locate the center lines to bring the drawing in the center of the rectangle.

INKING

Have the drawing complete in pencil before inking. When inking, follow the order given on page 94.

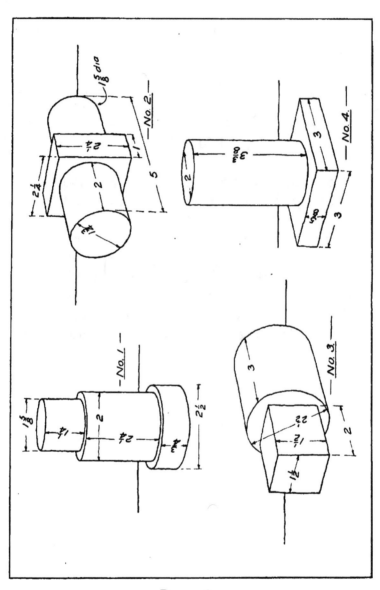

PLATE 9

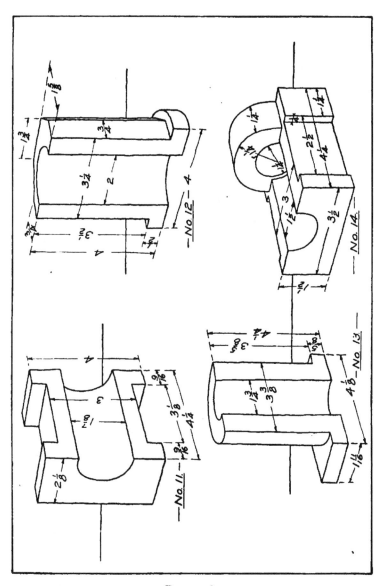

PLATE 9

EXTRA PLATE

Make two views of one of the objects shown on page 99.

Follow the general directions given for Plate 9.

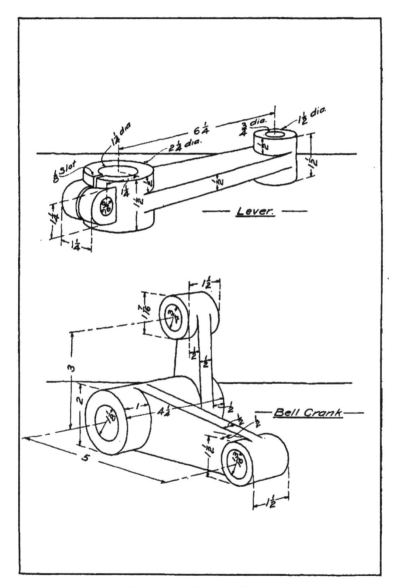

PLATE 9. EXTRA PLATE

PLATE 10

Copy the two views shown in one of the rectangles on pages 101 to 106, and work out the third view. The problem will occupy the entire sheet. Locate the center lines to bring the drawing in the center of the sheet. Print the name of the piece under the drawing, making the capitals $\frac{3}{16}''$ high and the small letters $\frac{1}{8}''$ high.

INKING

When inking, follow the order given on page 94.

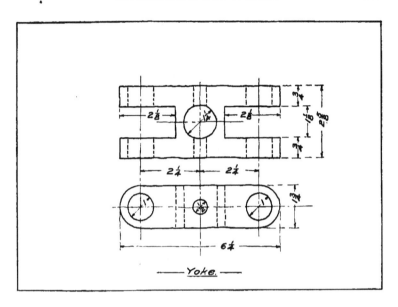

Yoke.

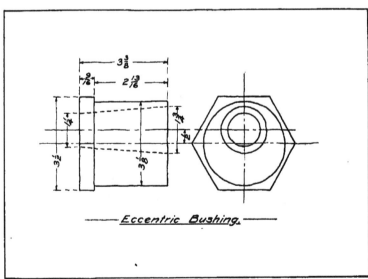

Eccentric Bushing.

PLATE 10

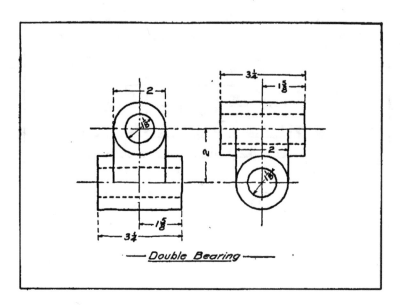

— Double Bearing —

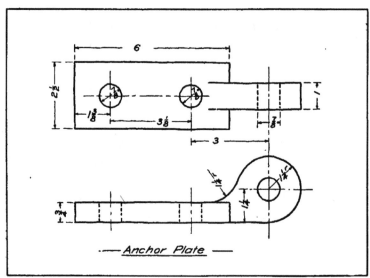

— Anchor Plate —

PLATE 10

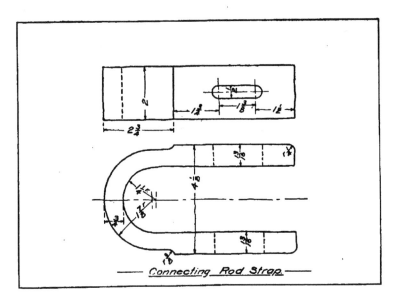

Connecting Rod Strap

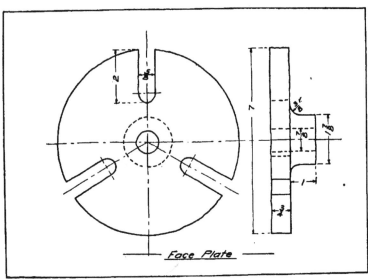

Face Plate

PLATE 10

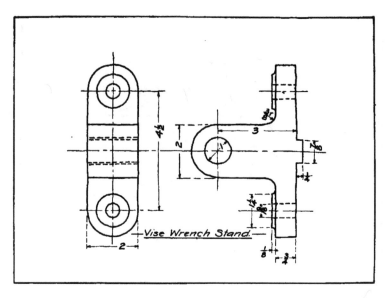

Vise Wrench Stand.

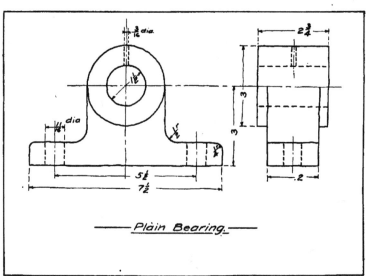

Plain Bearing.

PLATE 10

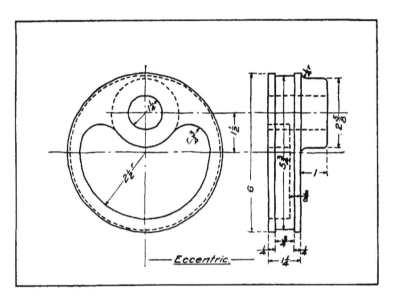

Eccentric.

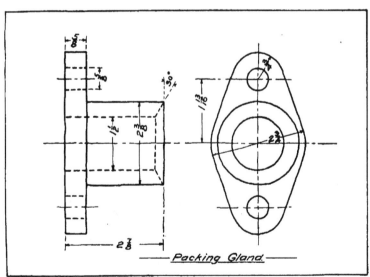

Packing Gland.

PLATE 10

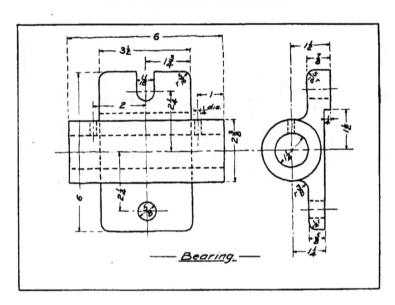

Bearing

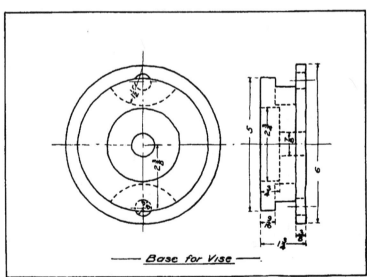

Base for Vise

PLATE 10

EXTRA PLATE

Copy the two views shown in one of the rectangles on pages 108 and 109, and work out the third view. The problem will occupy the entire sheet. Locate the center lines to bring the drawing in the center of the rectangle. Print the name of the piece below the drawing, making the capitals $\frac{3}{16}''$ high and the small letters $\frac{1}{8}''$ high.

INKING

When inking, follow the order given on page 94.

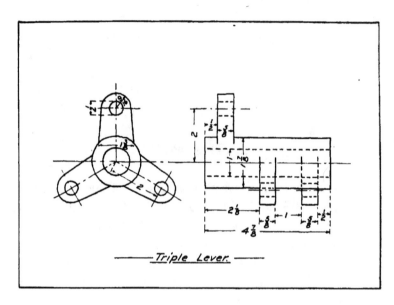

Triple Lever.

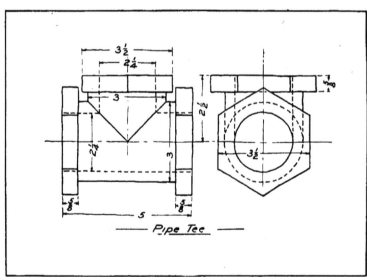

Pipe Tee

PLATE 10. EXTRA PLATE

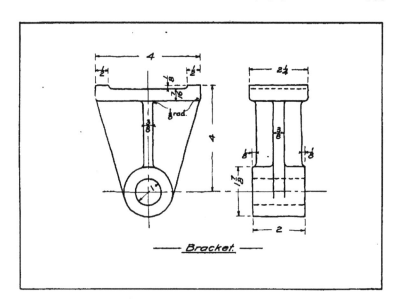

Bracket.

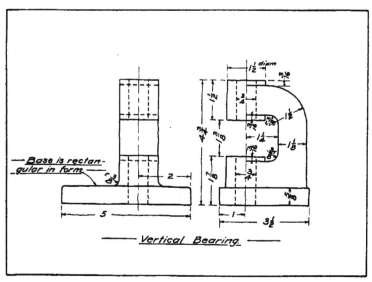

Base is rectangular in form.

Vertical Bearing.

PLATE 10.   EXTRA PLATE

CHAPTER XI

SCALED DRAWINGS

PLATE 11

Many times the object to be represented in the drawing is of such a size that it is impossible to draw it full size upon a sheet that may be conveniently handled in the shop. In cases of this kind, the drawing is made to a reduced scale, and is called a scaled drawing. A scaled drawing is one in which the length of all the lines of the drawing bears a definite ratio to the length of the corresponding lines of the object. Thus in a drawing made one-half size, each line of the drawing will be one-half the length of the corresponding line of the object.

Our United States standard system of measurements requires that the denominator of the fraction used in our scale shall be some multiple of two. The drawings, therefore, will have to be made $\frac{3}{4}$, $\frac{1}{2}$, $\frac{1}{4}$ or $\frac{1}{8}$ size. It will be readily seen that it would be practically impossible to measure most dimensions, if a two-thirds or one-sixth scale were employed. Drawings should always be made to the largest possible scale. If it is possible to use three-fourths size without crowding, do not make the drawing one-half size. It is, of course, easier for the workman to work from the full-sized drawing, and the smaller the scale the more difficulty there will be in reading. Full size dimensions should be placed on the drawing; not the measurements of the drawing, but the dimensions to which the object is to be made.

A statement giving the scale to which the drawing is made should be printed on each scaled drawing. There are two ways in which this may be expressed. The one commonly used on machine drawings is *Scale, ¾ Size*, or *Scale, ½ Size*. Because of the small scales employed, the architect uses the following: *Scale, 3″ = 1′*, or *⅜ in. = 1ft.*

PLATE 11

Copy the two views given in one of the rectangles on pages 113 to 116, and work out the third view. Make the drawing to the largest possible scale. Place the title and the scale to which the drawing is made below the drawing.

INKING

When inking, follow the order given on page 94.

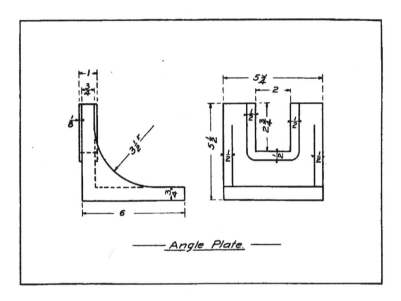

Angle Plate.

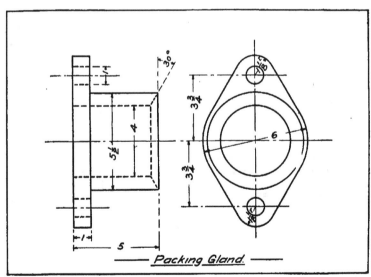

Packing Gland.

PLATE 11

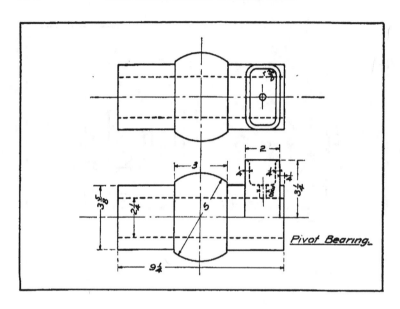

Pivot Bearing.

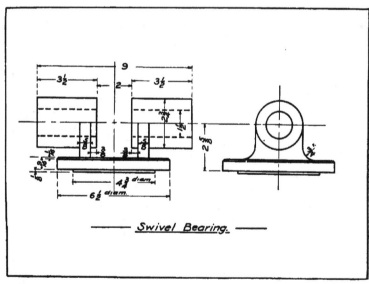

Swivel Bearing.

PLATE 11

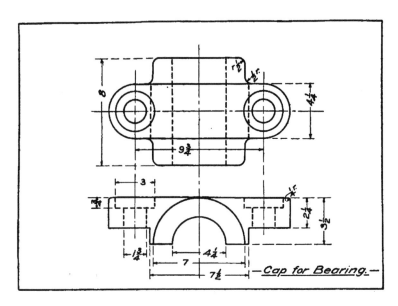

—Cap for Bearing.—

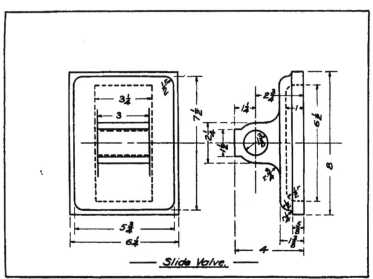

— Slide Valve. —

PLATE 11

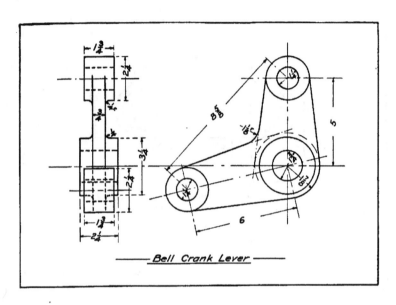

Bell Crank Lever

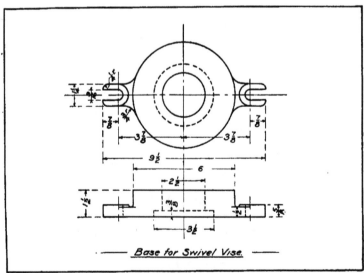

Base for Swivel Vise.

PLATE 11

EXTRA PLATE

Copy the two views given in one of the rectangles on page 118, and work out the third view. Make the drawing to the largest possible scale. Place the title and the scale to which the drawing is made below the drawing.

INKING

When inking, follow the order given on page 94.

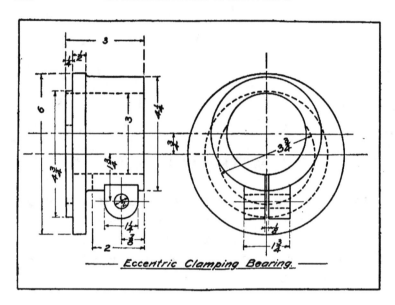

Eccentric Clamping Bearing.

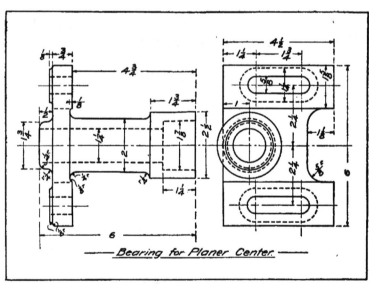

Bearing for Planer Center.

PLATE 11. EXTRA PLATE

CHAPTER XII

SECTIONAL VIEWS

PLATE 12

In the problems which have been presented thus far in the course, emphasis has been placed upon expression with full lines in preference to dotted ones. Dotted lines are confusing on any drawing, and should be avoided wherever possible. To eliminate the dotted lines caused by the three views we have already discussed, a system by which the interior of the object may be shown in full lines is employed. Views of this character are known as sectional views, and are obtained by passing an imaginary cutting plane through the object and making a drawing of the portion of the object remaining after the part cut off by the cutting plane has been removed.

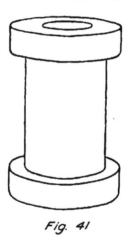

Fig. 41

A double flanged cylinder with a circular hole passing through it is shown in Fig. 41. Let us imagine that this cylinder is cut in halves by a vertical plane passing through the axis. After passing this cutting plane through the piece, and removing the front half of the cylinder, we would have what is represented

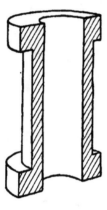

in Fig. 42. Two views of the model shown in perspective in Fig. 41 are given in Fig. 43, while in Fig. 44 are shown two views after the imaginary cutting plane has been passed through the piece. It will be seen that the front view only is altered, the top view remaining the same. The model represented is a complete cylinder, and not a half cylinder; therefore the complete top view is necessary. The front view, which really represents only one-half of the cylinder, is produced by the imaginary cutting plane, and not an actual cutting plane; hence the complete top view.

Fig. 42

A comparison of the full front view in Fig. 43 and the sectional front view in Fig. 44 brings out the fact that the hole through the piece is represented by dotted lines on the former and by full lines on the latter. Also note that portions of the horizontal lines representing the lower part of the upper flange and the upper part of the lower flange have been omitted on the sectional view. It is common practice to omit the dotted lines, when the representation is complete by full lines, even though every line of the object is not shown on the drawing. Had the flanges been square, as shown in Fig. 45, it would have been

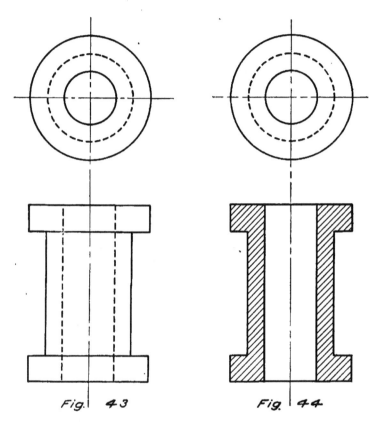

Fig. 43   Fig. 44

best to draw these dotted lines, in order to show the corner on the front view.

The oblique lines drawn across portions of the sectional views are called section lines. These are drawn only where the material of the model is cut by the cutting plane. In Figs. 44 and 45, they are not drawn across the area representing the hole through the piece, but only across those portions of the model where the cutting plane has come in contact with the material of which the model is made. The section line may be made at any angle and in either direction. The most common angle is 45°, and the direction is usually from the lower left to the upper right, probably because of its convenience. Section lines should not be drawn with pencil, nor should they be spaced with the instruments or scale. They should be inked with black ink, with a line finer than the finest line of the drawing, and should be spaced with the eye, care being taken not to have the lines too close together. For average work, spaces should be about $\frac{3}{32}''$.

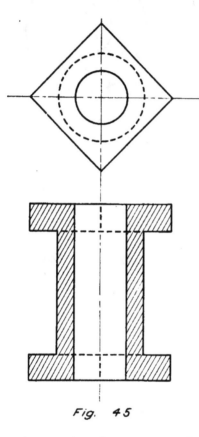

Fig. 45

When two or more pieces are in contact and a sectional view is used, the section lines for the several pieces should be drawn in opposite directions. Should these pieces be so placed that one of them is in contact with two or more, it will be necessary to change the angle of the lines as well as the direction.

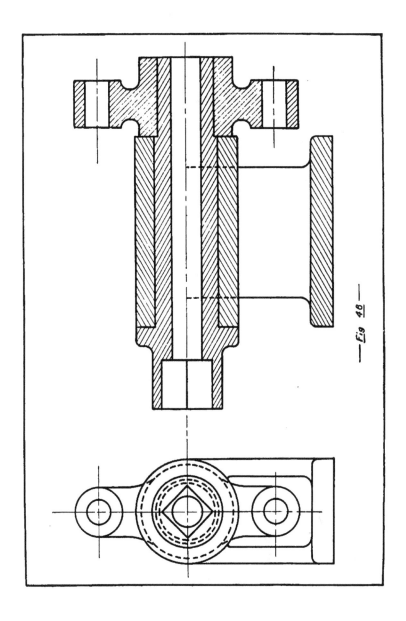

Fig 48

In Fig. 46, page 123, the inner piece or spindle is in contact not only with the bearing, but also with the yoke at the end of the piece. The section lines for the spindle are drawn at an angle of 30° with the horizontal, from the upper left to the lower right. For the bearing, the lines are drawn at an angle of 30° from the lower left to the upper right. The lines for the yoke are at an angle of 45°, and, though drawn in the same direction as the lines of the spindle, the difference is shown by the change in angle.

The introduction of the section lines causes another group to be added to those already classified for inking. In all of the work hereafter, use the order for inking given below.

ORDER FOR INKING

Group 1. Object lines; heavy lines, with black ink.
Arcs; begin with the largest.
Horizontal lines; the upper ones first.
Vertical lines; those at the left first.
Oblique lines; the most convenient way.

Group 2. Witness, center, and dimension lines; light lines, with red ink.
Arcs; begin with the largest.
Horizontal lines; the upper ones first.
Vertical lines; those at the left first.
Oblique lines; the most convenient way.

Group 3. Arrow-heads, dimensions, and printing; with black ink.
Free hand, with writing pen. Work from the upper left to the lower right.

Group 4. Section lines; very light, black lines. Finer than any other line of the drawing.

Group 5. Margin lines; heavy lines, with black ink. When these are made the same size as the object lines, they should be inked with Group 1.

EXTRA PLATE

Copy the two views shown in one of the rectangles on pages 133 and 134. Make one of the views a sectional view.

INKING

When inking, follow the order given on page 124.

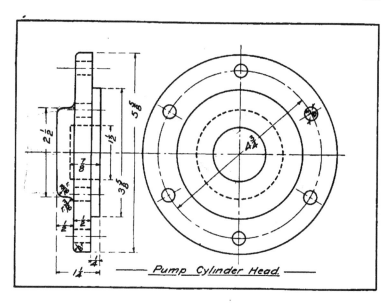

Pump Cylinder Head.

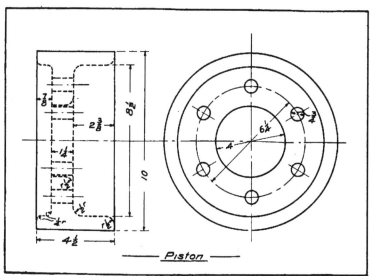

Piston

PLATE 12

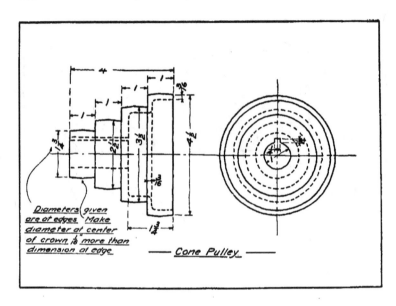

Diameters given
are at edges. Make
diameter at center
of crown $\frac{1}{8}$ more than
dimension at edge

— Cone Pulley —

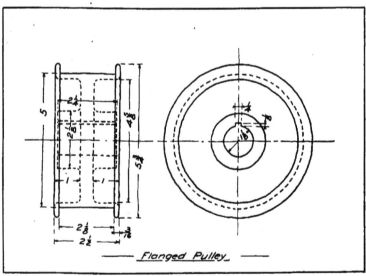

— Flanged Pulley —

PLATE 12. EXTRA PLATE

CHAPTER XIII

PARTIAL SECTIONS

PLATE 13

In order to show both the exterior and the interior in full lines in a single view, a half section, shown in Fig. 47, is often employed. This, it will be noted, is a drawing of the model shown in perspective in Fig. 41, and the front view is a combination of the front views given in Figs. 43 and 44, page 121, the right half being the same as the right part of the full view, while the left portion is the same as the left half of the sectional view. The cutting plane to produce this front view would pass along the horizontal center line of the top view to the center of the piece, then along the vertical center line to the front of the view. Denoting the path by letters, the cutting plane would pass along the line A B C. Thus it will be seen that the path of the cutting plane need not be a straight line, as was the case in Figs. 44 and 45. In Fig. 47 the place where the cutting plane ends is shown by a full object line, but it is often the practice to use simply the center line to limit this sectioned area. This method is shown in Fig. 48. In this view all of the dotted lines are omitted.

Another instance of a cutting plane not being con-

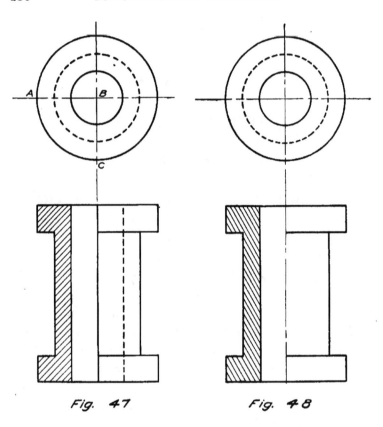

Fig. 47　　　　　Fig. 48

tinuous is shown in Fig. 49. In this example the cutting plane passes along the horizontal center line to the shaft upon which the disc is mounted, thence around the shaft to the horizontal center line again, and along this center to the outside of the piece. This leaves the spindle or shaft in full on the front view, which leads to the statement that *solid, cylindrical parts should never be shown in section.* The full length of the shaft is not shown in the front view and

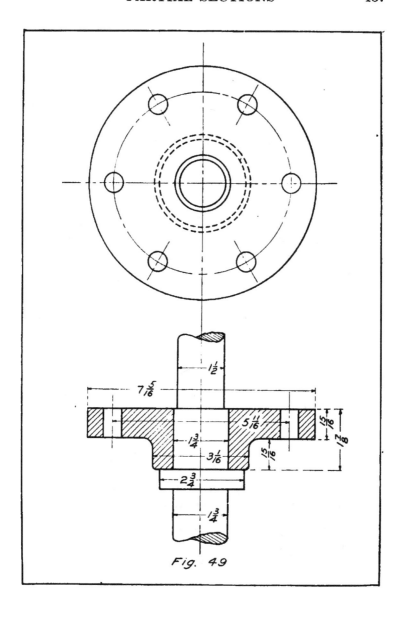

Fig. 49

ends at top and bottom in what are termed *broken sections*. This figure also shows the breaking of the section lines for the dimension, when the dimension has to be placed in the sectioned area. Note that the breaking of the section lines brings into prominence the $5\frac{11}{16}''$ dimension on the front view.

Arms of pulleys and hand-wheels, ribs on castings, etc., should be placed back of the cutting plane, when a sectional view is used. The application of this statement to the arms of a pulley is shown in Fig. 50. The line of the cutting plane to produce the sectional view would be along the vertical center line of the side view. This would make the arm of the pulley in section; but, in cases of this character, we draw the sectional view as though the plane passes through the rim and hub, leaving the arm as a full view. By this method one is able to distinguish, by a glance at the sectional view, the character of the connection between the hub and rim of a pulley. A pulley with a web connection between rim and hub is illustrated in Fig. 51, and a comparison of the two sectional views will show the way in which one may differentiate between the two designs. Had the sectional view of the arm pulley been made by passing the cutting plane through the arm, the resultant view would have been the same as the sectional view of the webbed pulley, with no opportunity to distinguish between the two, except by reference to the other view. By placing the arm back of the plane of the section, we emphasize the character of the design.

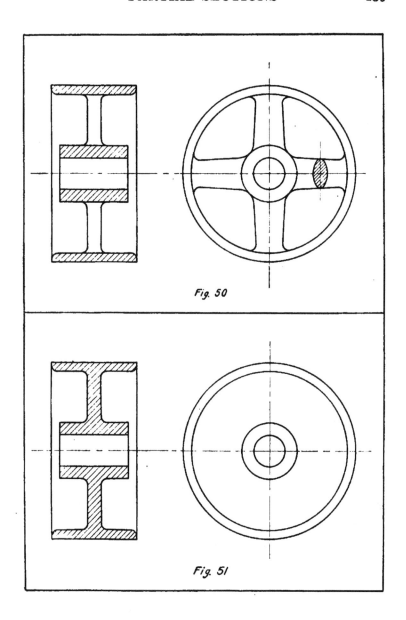

Fig. 50

Fig. 51

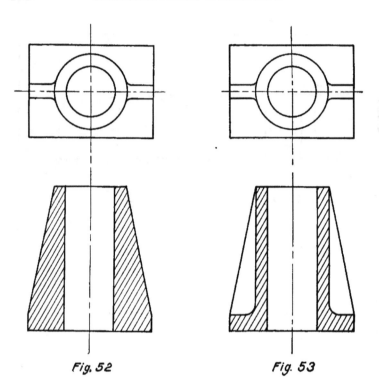

Fig. 52   Fig. 53

The sectional view of the model shown in Fig. 52 gives the impression of a conical piece, while in reality it is a cylindrical piece with supporting ribs and rectangular base.   While it is possible to see the exact shape by the combination of the front and top views, yet no one view should give a wrong impression.   If, instead of the views given in Fig. 52, we had used the views shown in Fig. 53, the sectional view would give the correct impression of the general shape of the object.   This would, then, be the proper view to employ.   It is obtained by placing the ribs back of the plane of the section, as we did the arms of the pulley in a previous illustration. This drawing also shows the size of the fillet where the cylinder joins the base, something which was not shown at all in Fig. 52.

From the drawing given in Fig. 50, it is impossible to tell the shape of the arm of the pulley. We can

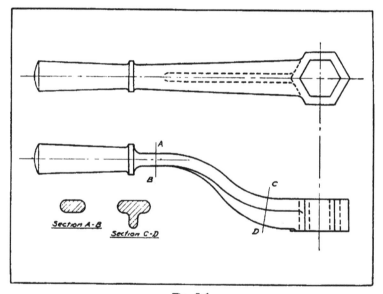

Fig. 54

get the width and thickness, but whether it is rectangular, flat on the sides with rounded corners, or elliptical in section, we do not know. The section drawn on the arm at the right of the center shows the cross section to be an ellipse. Here the view is shown at right angles to the cutting plane, a practice quite common in cases of this character. Often the line of the cutting plane is given and the section drawn in another place. This is illustrated in Fig. 54. In this figure the lines A B and C D represent the lines of two cutting planes, and the figures below the drawing show the shapes of the sections. It is necessary to distinguish them by letters and notes, as shown in the figure.

PLATE 13

Copy the drawing given in one of the rectangles on pages 144 to 146, making the upper half of one of the views in section.

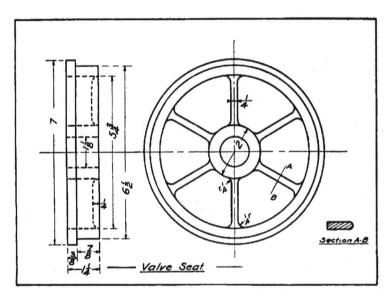

Valve Seat.

Section A-B

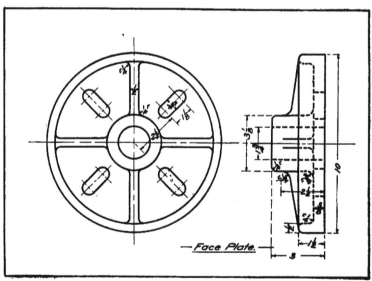

—Face Plate.—

PLATE 13

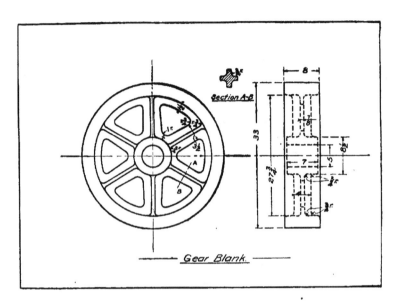

Gear Blank.

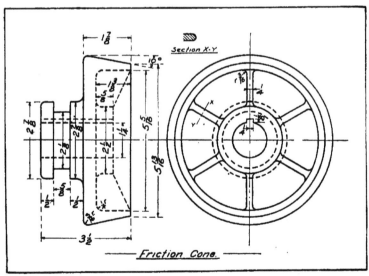

Friction Cone.

PLATE 13

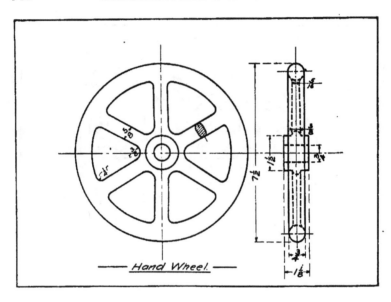

— Hand Wheel. —

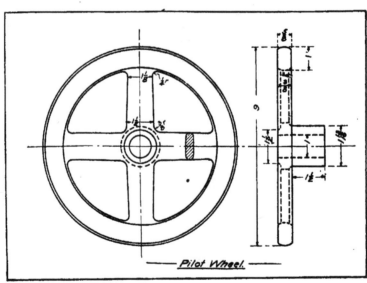

— Pilot Wheel. —

PLATE 13'

EXTRA PLATE

Follow the directions given in one of the rectangles on pages 148 to 150.

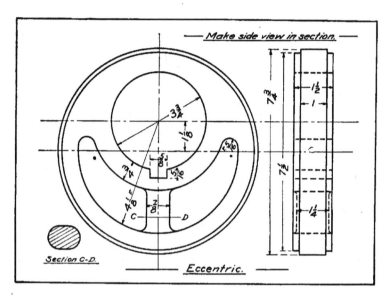

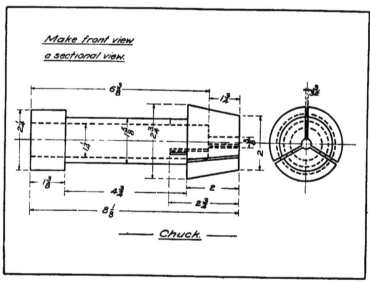

PLATE 13. EXTRA PLATE

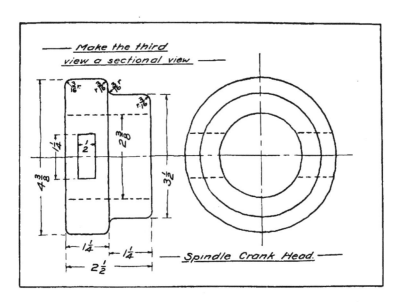

Make the third
view a sectional view —

Spindle Crank Head.

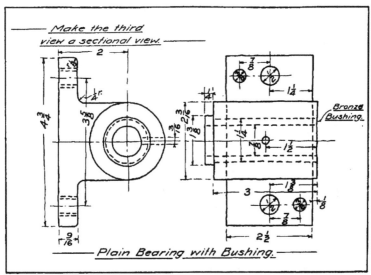

Make the third
view a sectional view. —

Bronze Bushing

Plain Bearing with Bushing.

PLATE 13. EXTRA PLATE

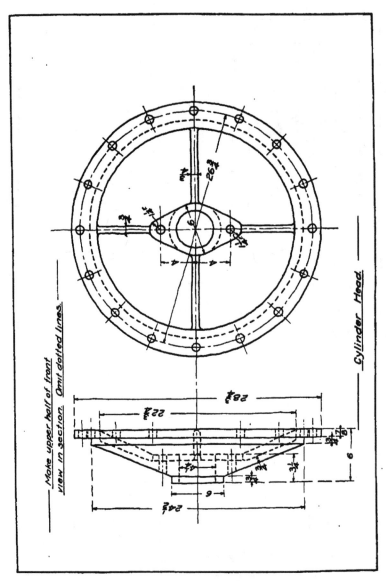

PLATE 13. EXTRA PLATE

CHAPTER XIV

LETTERING

CAPITALS

Good lettering is an essential to good work in mechanical drawing. Plain, simple, well-proportioned, properly spaced lettering will improve the appearance of any drawing, no matter how beautifully it may be executed otherwise; and it is equally true that the appearance of the drawing will be ruined by poor lettering.

On almost all mechanical drawings, the Gothic type, an example of which is shown on page 154, is employed. Sometimes all capitals are used, but in the majority of cases the capitals and small letters will be found. When all capitals are used, the capitalized letters are made higher than the others. (See Fig. 55.) A careful examination of the alphabet will show that the letters are composed of either straight lines, ellipses, or combinations of the two. For instance, the capitals H, N, M and others are composed entirely of straight lines; the O and Q are ellipses; C and G are portions of ellipses; while D, P, R, and others, are combinations of parts of ellipses and straight lines. Combinations of a similar nature will be found in the small letters.

ABCDEFGHIJKL

MNOPQRSTU

VWXYZ

abcdefghijklmno

pqrstuvwxyz

1234567890

$4\frac{1}{4}$ $2\frac{1}{2}$ $7\frac{3}{4}$

The main body of the small letter should be two-thirds the height of the capital. A good size for practice work, in fact, for most lettering on a drawing, is $\frac{3}{16}$ of an inch for capitals and $\frac{1}{8}$ of an inch for the main part of the small letter. Lines should always be drawn regulating the height, and the letters should exactly fill the space between the lines.

Particular care should be used to *make the slant of the letters uniform*. This slant may be made anything between 60° and 75° with the horizontal, the latter being the better angle. In the capitals A, V, X, and W, and the small letters v, x, and w, the angle of the component parts may be determined by the use of a center line. In the word "Vanity," Fig. 55, the general slant is shown by the letters I, N, and T, and the slope of the parts of the letters V, A, and Y is obtained by laying off equal spaces each side of center lines drawn parallel to the letter I.

The manner of obtaining the slant of the curved letters may best be illustrated by using O· as an example. Draw a parallelogram (see Fig. 56) in which the upper and lower lines shall limit the height of the letters, and the right and left-hand lines shall represent the slant. Divide each side into two equal parts. This gives the tangent points of an ellipse to be drawn within the parallelogram.

It should be perfectly understood that all of this work, with the exception of the lines to regulate the height of the letter, should be executed free hand,

and that as soon as the student can trust himself, all of the construction for the curved letters should be omitted. When inking, only a small amount of ink should be used on the pen, and great care should be exercised not to spread the points. Avoid all shading and shade line effects.

No definite rule can be given for the space between letters, this space varying with different combinations. The space between words should be about three times the average space between letters.

SUGGESTED EXERCISES FOR HOME WORK

Capital Letters

(Make the capital letters $\frac{3}{16}''$ high)

1. A series of parallel oblique lines about $\frac{5}{16}''$ apart.

2. A series the same as No. 1, with the addition of the horizontal line to form the letter L.

3. The same as No. 1, with the additional line to make the letter T.

4. A series of parallel lines having alternate small and large spaces.

Make the small spaces about $\frac{5}{32}''$ and the large ones about $\frac{5}{16}''$.

5.

A series similar to No. 4, with the addition of the cross line in the small spaces to form the letter H. Notice that the cross bar is slightly above the center.

6.

·The same as No. 4, with the additional line to make the letter N.

7.

The same as No. 1, with additions to form the letter E. Note that the upper horizontal line is shorter than the lower one, and that the middle one is slightly above the center.

8.

The same as No. 1, with additions to form the letter F. This is the letter E with the lower line omitted.

9. K K K K

The same as No. 1, with the lines to form the letter K.

10. M M M M

The width of the M is equal to the height. Either form may be used, although the one at the right is preferable.

First draw a series of center lines about $\frac{7}{16}''$ apart. About these center lines construct the

11. A A A A

letters. After becoming familiar with the relation of the component parts, draw a series without the use of the center lines.

12. V V V V

Follow the general directions given for No. 11.

The center lines of the two parts of

13. W W W W

the W should be about $\frac{1}{8}''$ apart. Having drawn the center lines, con-

struct the letter.   Finally, draw the letters without
the center lines.

14. X X X X

15. Y Y Y Y

16. Z Z Z Z

17. O O O O

Draw a series of
center lines as di-
rected for No. 11.
The intersection of
the c r o s s lines
should be slightly
above the center of
the letter.

The center line
of the letter forms
the lower part of
the letter Y. The
vertex of the V is
slightly below the
center of the letter.

Draw the con-
struction lines as
directed for No. 4.
Form the letter by
the use of these
lines. Finally draw
the letter without
the use of the con-
structions.

Draw the con-
struction lines to
the directions given
for No. 4. Read
the text given in

connection with Fig. 56, page 155. Follow the directions there given. Finally, make the letter without the use of the construction lines.

The same constructions are used as for the letter O. The C is the letter O with a portion at the right left out. The G is the letter C with the additions shown.

18.

The combination of the lower part of the letter O with straight lines forms the letter U. Use the same constructions as for the letter O, No. 17.

19.

The right hand portion of the letter D is the same as the corresponding part of the letter O.

20.

Note that the loop of the letter P is less than one-half of the height of the letter. The

21.

curve is one-half of an O, joining horizontal lines. The R is the letter P with an additional line. Use spacing given for No. 1.

Note in the letter B that the upper loop is smaller than the lower one. This letter is composed of curves similar to the letter O, joined to horizontal lines. Use the spacing given for No. 1.

22. B B B B

The letter S, probably the most difficult of all the letters to make, is one continuous curve. Considerable practice will be required to make this letter well. Use the spacing given for No. 4.

23. S S S S

SMALL LETTERS

Some of the small letters are exactly the same shape and have the same proportions as the capitals, the only way of distinguishing them from the capitals

being by the height. A reference to page 154 will
show that this is true of the letters c, o, s, v, w, x, and
z. The letter l is a straight line, resembling the
capital I. The main lines of the t and i are also
straight lines. The small u is similar to the capital
U, only the straight line at the right continues down
below the loop. The letters f, j, k, and r are easily
formed, and require no special directions. The
letters a, b, d, g, p, and q are based upon the letter
o and a straight line tangent. The upper parts of
the loops of the h, m, and n is the curve of the upper
part of the letter o. From this it will be seen that a
perfect mastery of the curves of the letter o is absolutely necessary.

SUGGESTED EXERCISES FOR HOME WORK

Small Letters

(Make the body of the small letters ⅛″ high)

24.

The letter o until perfectly mastered. Make the letters ⅛″ and have a space of about ₁₆³″ between each letter.

25.

The letter o with the tangent straight line forming the letter a. Many think it better to draw the straight line first.

The letters b and d consist of the letter o with the straight line tangent. Notice that the d is shorter than the b.

The letter o with the straight line tangents extending below the line forms the p and q. Notice the slight curve at the lower part of the straight line of the q.

The small e is the letter c with the horizontal line addition. It is best to draw the horizontal line last.

The loops of the h and n are the same as the upper part of the letter o. The remaining parts of the letters are made up of straight lines.

26.

27.

28.

29.

30.   The m is similar
to the n, only the
l o o p s are nar-
rower.

FIGURES

Copy the series of figures given on page 154, until a perfect mastery has been obtained.

CHAPTER XV

GEOMETRICAL DEFINITIONS

POINT

A *point* indicates position only, and has no dimension.

LINES

A *line* is produced by the motion of a point and has dimension in length only. (In drawing a line with a pencil, the successive positions of the pencil point produce the line.)

Lines may be straight, curved, broken or mixed.

A *straight line* is a line which has the same direction throughout its entire length. See Fig. 57. A straight line is frequently called a *right line*.

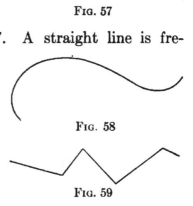

FIG. 57

A *curved line* is a line no part of which is straight. See Fig. 58.

FIG. 58

A *broken line* is a series of straight lines drawn in different directions. See Fig. 59.

FIG. 59

A *mixed line* is a line composed of straight and curved lines. See Fig. 60.

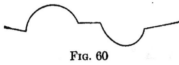

FIG. 60

Straight lines are often called simply *lines;* and curved lines, *curves.*

Straight lines may be *horizontal, vertical* or *oblique.* A line drawn from left to right is termed a *horizontal* line. A horizontal line is shown at A B, Fig. 61.

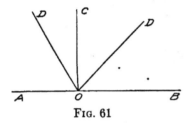

FIG. 61

Lines drawn in the opposite direction are called *vertical* lines. Line O C, Fig. 61, is a vertical line.

Any straight line neither horizontal nor vertical is called an *oblique* line. See lines D O, Fig. 61.

Parallel lines are everywhere equally distant from each other. The lines shown in Fig. 62, are parallel.

FIG. 62

A line is *perpendicular* to another line when it meets that line so as not to incline towards it on either

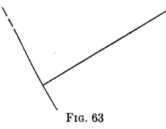

FIG. 63

side. (When speaking of perpendicular lines, the relation of one line to another is always understood. Thus, a vertical line when alone is not a perpendicular and is only spoken of as such when it is referred to in connection with a horizontal line. The horizontal and vertical lines in Fig. 61 are perpendicular, which is equally true of the two lines shown in Fig. 63.)

ANGLES

The opening between two straight lines which meet is called an *angle.* The lines M R and M N, Fig. 64, form an angle. The lines are called the *sides* of the angle, and the point of meeting is known as the *vertex.*

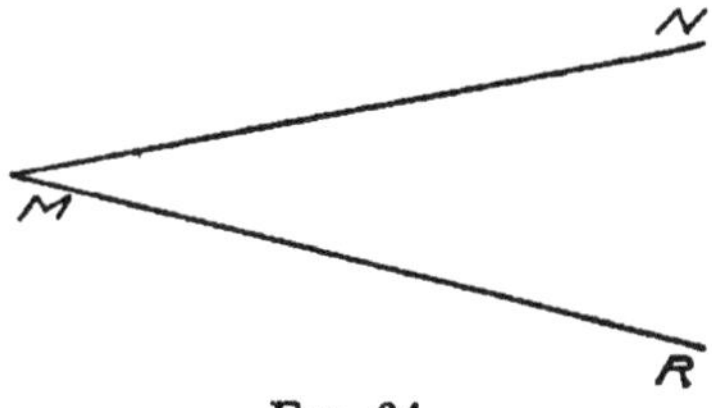

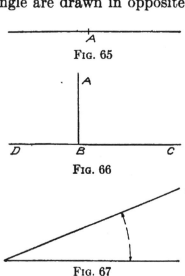

FIG. 64

The *size of an angle* depends, not upon the length of its sides, but upon the amount of the opening between the sides.

When the sides of an angle are drawn in opposite directions they form a *straight angle.* L i n e s drawn in opposite directions from the point A, Fig. 65, form a straight angle.

FIG. 65

A *right angle* is formed when the sides are perpendicular. The angles A B C and A B D, Fig. 66, are right angles.

FIG. 66

Every angle less than a right angle is an *acute angle.* See Fig. 67.

FIG. 67

When an angle is greater than a right angle and less than a straight angle, the angle formed is called an *obtuse angle*. See Fig. 68.

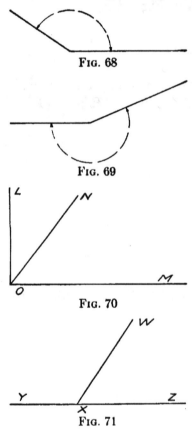

FIG. 68

FIG. 69

FIG. 70

FIG. 71

An angle greater than a straight angle and less than two straight angles is called a *reflex angle*. Fig. 69 represents a reflex angle.

Two angles are *complementary* when their sum is equal to a right angle. Each is called the complement of the other. In Fig. 70, the angles L O N and M O N are complementary.

When the sum of two angles is equal to a straight angle, the angles are called *supplementary*, and each is called the supplement of the other. The angles W X Y and W X Z, Fig. 71, are supplementary.

TRIANGLES

A *triangle* is a plane surface bounded by three straight lines. The bounding lines are called the *sides* of the triangle, the angles formed by the sides

are called the *angles* of the triangle and the vertices
of these angles are the *vertices* of the triangle.   The

base of a triangle is the
side upon which it is supposed to stand.   The
angle opposite the base
is known as the *vertical
angle*, and the vertex of
this angle is called the
vertex of the triangle.

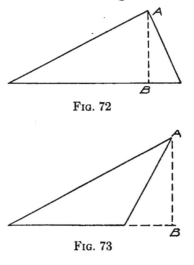

FIG. 72

The *altitude* of a tri-
angle is the perpendicu-
lar distance from the
vertex to the base or the
base produced.   See line
A B, Figs. 72 and 73.

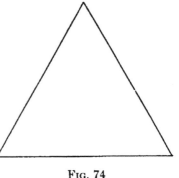

FIG. 73

Triangles classified by sides

An *equilateral triangle* is one having all of its sides

equal in length.   See
Fig. 74.   This triangle
is frequently called equiangular, as all of its
angles are also equal.

An *isosceles triangle* is
one having two of its
sides equal.   See Fig. 75.
Two of the angles of an
isosceles triangle are also
equal.

A *scalene triangle* is one having no two of its sides

FIG. 74

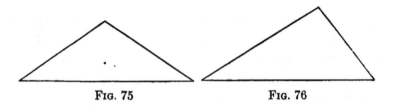

FIG. 75   FIG. 76

equal. See Fig. 76. None of the angles of a scalene triangle are equal.

Triangles classified by their angles

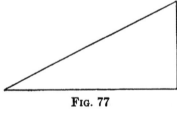

FIG. 77

A *right-angled triangle* or *right triangle* is one having a right angle. See Fig. 77.

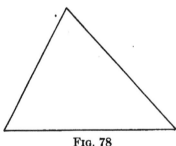

FIG. 78

An *acute-angled triangle* is one having three acute angles. See Fig. 78.

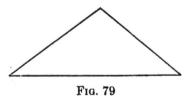

FIG. 79

An *obtuse-angled triangle* is one having an obtuse angle. See Fig. 79.

QUADRILATERAL

A *quadrilateral* is a plane surface bounded by four straight lines.

A *trapezium* is a quadrilateral which has no two of its sides parallel. See Fig. 80.

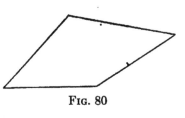

FIG. 80

A *trapezoid* is a quadrilateral having two and only two of its sides parallel. See Fig. 81.

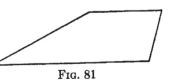

FIG. 81

A *parallelogram* is a quadrilateral having its opposite sides parallel. There are four kinds of parallelograms: the rectangle, the square, the rhomboid, and the rhombus.

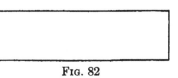

FIG. 82

A *rectangle* is a parallelogram whose angles are all right angles. Rectangles are shown in Figs. 82 and 83.

A *square* is a rectangle all of whose sides are equal. See Fig. 83.

FIG. 83

A *rhomboid* is a parallelogram whose angles are not right angles. Figs. 84 and 85 represent rhomboids.

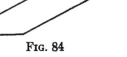

FIG. 84

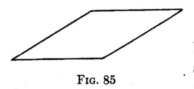

FIG. 85

A *rhombus* is a rhomboid all of whose sides are equal. See Fig. 85.

POLYGONS

A *polygon* is a plane surface bounded by straight lines. This term is usually applied to figures having more than four sides. The number of sides determines the name of the polygon.

A polygon of five sides is called a *pentagon;* one of six sides is a *hexagon,* one of seven sides is a *heptagon,* one of eight sides is an *octagon,* one of nine sides is a *nonagon,* and one of ten sides is a *decagon.*

A *regular polygon* has all of its sides and all of its angles equal.

A polygon is irregular when either sides or angles are unequal.

CIRCLE

A *circle* is a plane figure bounded by a curved line called the *circumference,* every point of which is equidistant from a point within called the *center.* See Fig. 86.

The *radius* of a circle is a straight line drawn from the center to a point in the cir-

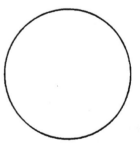

FIG. 86

cumference.   Line A B, Fig. 87, is a radius.   All
the radii of a circle are equal.

The *diameter* of a circle is a straight line drawn
through the center and
joining two points in the
circumference.   Line C
D, Fig. 87, is a diameter.
A diameter is equal to
two radii.

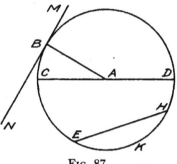

FIG. 87

An *arc* of a circle is any
part of its circumference,
as E K H, Fig. 87.   An
arc equal to one-half of the
circumference is called a *semicircumference*.   The arc
C K D, Fig. 87, is a *semicircumference*.

A *chord* is a straight line joining the extremities
of an arc.   See line E H, Fig. 87.

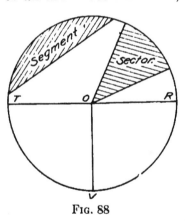

FIG. 88

A *tangent* is a straight
line which touches the circumference of a circle but
does not intersect it, as
M N, Fig. 87.   The point
at which the tangent
touches the circle is called
t h e *point of tangency.*
The tangent is always
perpendicular to the ra-
dius at the point of tan-
gency.

A *segment* of a circle is the area bounded by an arc
and its chord.   See Fig. 88.   A segment equal to

one-half the circle is called a *semicircle*. See R V T O, Fig. 88.

A *sector* is the area bounded by two radii and the arc which they meet. See Fig. 88. When the radii are perpendicular, the sector equals one-fourth of a circle and is called a *quadrant*. See R O V, Fig. 88.

Every circle is supposed to be divided into 360 parts, called *degrees*, which are used as a measurement for angles. An arc of a semicircle, or straight angle, is equal to 180 degrees. An arc of a quadrant, or right angle, is equal to 90 degrees.

SOLIDS

A *polyhedron* is a solid bounded by planes. The bounding planes are the *faces*, and their intersections are the *edges* of the polyhedron.

Polyhedrons are classified according to the shape and relation of their faces.

Prisms

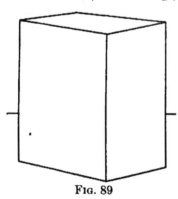

FIG. 89

A *prism* is a polyhedron of which two opposite parallel faces, called *bases*, are equal and parallel polygons, and the other faces, called *lateral faces*, are parallelograms. See Figs. 89 and 90. The intersections of the lateral faces of a prism are called *lateral edges*.

The *altitude* of a prism is the perpendicular distance between the bases.

A *right prism* is one whose lateral edges are perpendicular to the bases.   See Fig. 89.

A *regular prism* is a right prism whose bases are regular polygons.

An *oblique prism* is one whose lateral edges are not perpendicular to the bases. See Fig. 90.

A *truncated prism* is the part remaining between the base and a cutting

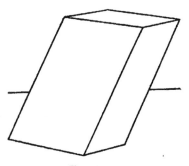

FIG. 90

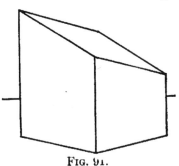

FIG. 91.

plane oblique to the base which intersects all of the lateral edges.   A truncated prism is shown in Fig. 91.

Prisms are named by their bases.   They are triangular, square, rectangular, hexagonal, etc., as the bases are triangles, square, rectangles, hexagons, etc.

Pyramids

A *pyramid* is a polyhedron one face of which, called the *base*, is a polygon and whose lateral faces are triangles whose vertices meet in a common point called the *vertex of the pyramid*.   See Fig. 92.

The *altitude* of a pyramid is the perpendicular distance between the base and the vertex.

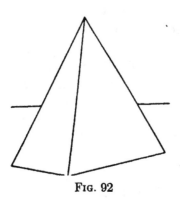

Fig. 92

A *regular pyramid* is a pyramid whose base is a regular polygon the center of which is in a perpendicular to the base let fall from the vertex.

A pyramid is triangular, pentagonal, octogonal, etc., as its base is a triangle, a pentagon, an octagon, etc.

The *frustrum of a pyramid* is the portion remaining between the base and a cutting plane parallel to the base which cuts all of the lateral edges. See Fig. 93.

Fig. 93

The *altitude of the frustrum* of a pyramid is the perpendicular distance between the base and the cutting plane parallel to the base.

Cylinders

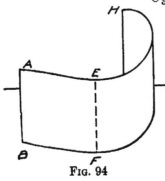

Fig. 94

A *cylindrical surface* is a curved surface generated by a moving straight line which constantly touches a given curve, and moves so that any two positions are parallel. See Fig. 94. In this figure, if the line A B moves parallel to the position lettered and is constrained to follow the curve A E H, a cylindrical surface is produced. Any position of this moving line, E F, parallel to A B, Fig. 94, is called an *element* of the surface.

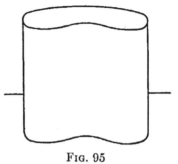

FIG. 95

A *cylinder* is a solid bounded by a cylindrical surface and two parallel planes, called *bases*. See Figs. 95 and 96.

A *right cylinder* is one whose elements are perpendicular to the bases. See Figs. 95 and 96.

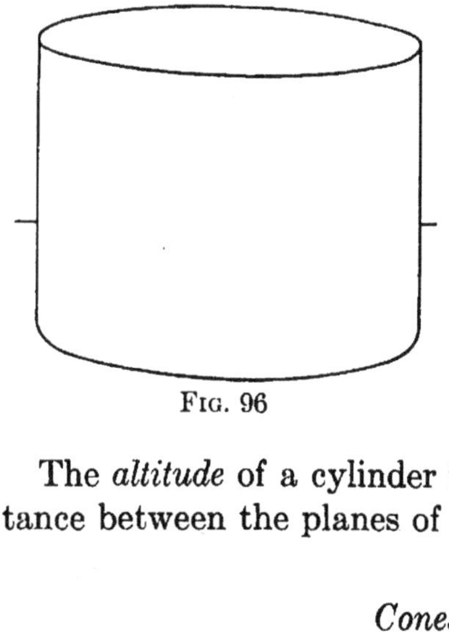

When the elements of the cylindrical surface are not at right angles to the bases, the cylinder is called an *oblique cylinder*.

A *circular cylinder* is a cylinder whose bases are circles. See Fig. 96.

FIG. 96

The *altitude* of a cylinder is the perpendicular distance between the planes of its bases.

Cones

A *conical surface* is a curved surface generated by a moving straight line one point of which is fixed while the line is made to follow a given curve. In Fig. 97, the line A B in its several positions passes through the

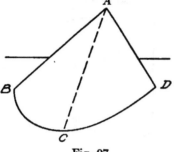

A

B

D

C

Fig. 97

fixed point A at the same time touching the curve B C D.

Any position of the moving line is called an *element* of the conical surface. Lines A B, A C, and A D, Fig. 97, are elements of the surface.

The fixed point through which the elements pass is called the *vertex*.

A *cone* is a solid bounded by a conical surface and a plane surface cutting all the elements of the conical surface. See Fig. 98.

The *altitude* of a cone is the perpendicular distance between the vertex and the plane of the base.

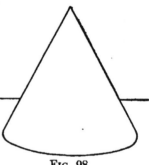

Fig. 98

A *circular cone* is one whose base is a circle.

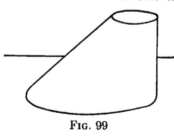

Fig. 99

A *right circular cone* is a circular cone whose vertex lies in a line drawn perpendicular to the plane of the base from its center.

The *frustrum of a cone* is the part contained between the base and a cutting plane parallel to the base. See Fig. 99.

The *altitude of a frustrum* is the perpendicular distance between the base and the cutting plane parallel to the base.

Sphere

A *sphere* is a solid bounded by a curved surface every point of which is equidistant from a point within called the center.

The *radius of a sphere* is the straight line drawn from the center to the bounding surface. All radii are equal.

The *diameter of a sphere* is a straight line drawn through the center and terminating in the spherical surface. The diameter is equal to two radii.

CHAPTER XVI

GEOMETRICAL PROBLEMS

In the figures accompanying these problems, the given lines are made heavy and full, the required lines are made heavy with a long and short dash, and the construction lines are made full, light lines.

PROBLEM 1

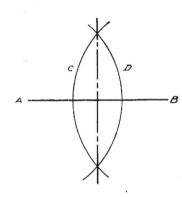

To bisect a straight line —With the ends A and B as centers and a radius greater than one-half the length of the line, draw arcs C and D intersecting on each side of the line. A line drawn through these intersections will bisect and be perpendicular to the given line.

PROBLEM 2

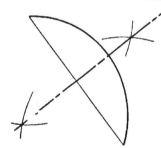

To bisect an arc—Draw the chord of the arc and bisect this chord. This bisector will bisect the arc and will pass through the center about which the arc is drawn.

PROBLEM 3

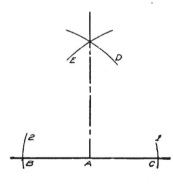

To draw a perpendicular to a line from a point near the center of the line— Using the given point A as a center, with any radius draw arcs 1 and 2, cutting the given line at B and C. With these points of intersection as centers, draw the arcs D and E.   A line drawn from the point of intersection of these arcs to the point A will be perpendicular to the given line.

PROBLEM 4

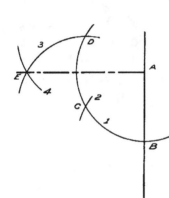

To draw a perpendicular to a line from a point at or near the end of the line—With any radius and the given point A as a center, draw arc intersecting the given line at B. With B as a center and the same radius, draw arc 2 cutting arc 1 at the point C. With C as a center and the same radius, draw arc 3, cutting arc 1 at D. With D as a center and the same radius, draw arc 4 cutting arc 3 at E.   A line drawn from E to A will be perpendicular to the given line.

PROBLEM 5

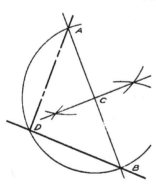

To draw a perpendicular to a line from a point outside of and near the end of the line —Through the given point A draw any line, such as A B, intersecting the given line. Find the point C by bisecting the line A B. (See Problem 1.)   Draw the semi-

circle A B D with C as a center. A line connecting
A and D will be perpendicular to the given line.

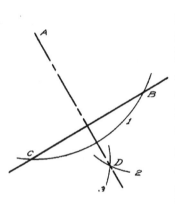

**To draw a perpendicular to
a line from a point outside of
and near the center of the
line**—With the given point
A as a center with any radius, draw arc 1, intersecting the given line at B and
C. With points B and C
as centers and equal radii,
draw arcs 2 and 3, inter-
secting at D.   The line A
D is the required perpen-
dicular.

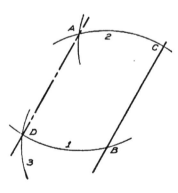

**To draw a line through a
given point parallel to a
given line**—With the given
point A as a center with
any radius, draw arc 1, intersecting the given line at
B. With B as a center with
the same radius, draw arc
2 through A intersecting
the given line at C.   With
a radius equal the distance

C A and B as a center, draw arc 3, intersecting arc 1 at D. A line drawn through A and D will be parallel to the given line.

PROBLEM 8

To divide a straight line into any number of equal parts— At any angle with the given line, A B, draw an indefinite line A C. Lay off on this line the required number of equal spaces. Through the points thus obtained draw a series of lines parallel to a line connecting the last point and the end of the line to be divided. (See Problem 7.) In the accompanying figure, the line A B is to be divided into five equal parts. On A C five equal spaces are laid off. Lines drawn through points 1, 2, 3, and 4 parallel to a line connecting B and 5 will divide the line A B into five equal parts.

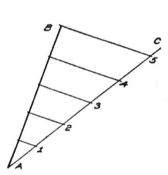

PROBLEM 9

Upon a straight line to construct an angle equal to a

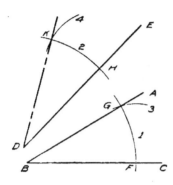

given angle—Let A B C be the given angle and D E the given line. With B as center and any radius, draw arc 1, cutting the side of the angle at F and G. With D as a center and with the same radius, draw arc 2, cutting D E at H. With F as a center, draw arc 3 through G. With the same radius and with H as a center, draw arc 4 cutting arc 2 at K. Angle K D H will equal angle A B C.

PROBLEM 10

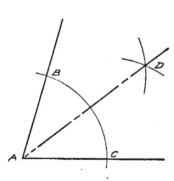

To bisect an angle—With A as a center and any radius, draw an arc, intersecting the sides of the angle at B and C. With B and C as centers and equal radii, draw arcs, intersecting at D. Line D A will bisect the angle B A C.

PROBLEM 11

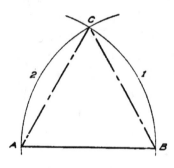

To construct an equilateral triangle on a given base,— The line A B is the given base. With A as a center, draw arc 1 through B. With B as a center, draw arc 2 through A and intersecting arc 1 at C. Lines C A and C B complete the triangle.

PROBLEM 12

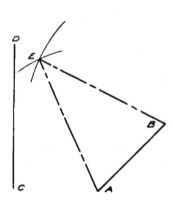

To construct an isosceles triangle on a given base, having given the length of the equal sides—The line A B is the given base and C D is the length of the equal sides. With the ends of the line A B as centers and a radius equal to the length of the line C D, draw arcs intersecting at E. The lines A E and B E complete the required isosceles triangle.

PROBLEM 13

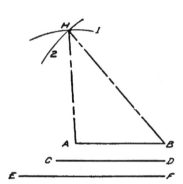

To construct a scalene triangle, the length of the sides being given—Let A B, C D, and E F be the length of the sides. With A as a center and the length of the side C D as a radius, draw arc 1. With B as a center and a radius equal to the length of the side E F, draw arc 2, intersecting arc 1 at H. Lines H A and H B complete the required scalene triangle.

PROBLEM 14

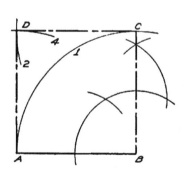

To construct a square, the length of the sides being given—The line A B is the length of one side of the square. Erect a perpendicular to A B at the point B. (Problem 4.) With B as a center, draw an arc through A, cutting line B C at C. With A and C as centers and the same radius, draw arcs 4 and 2 intersecting at D. Lines A D and C D complete the required square.

PROBLEM 15

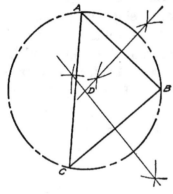

To circumscribe a circle about a triangle—Bisect two of the sides of the triangle, as A B and B C. (Problem 1.) With the intersection of these bisectors, point D, as a center, draw an arc through point A. This will be an arc of a circle passing through A, B, and C.

PROBLEM 16

To circumscribe a circle about a square—Draw the diagonals A C and B D intersecting at E. With E as a center and radius E B, draw the circumscribing circle.

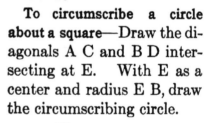

PROBLEM 17

To inscribe a circle within a triangle—Bisect two of the angles. (Problem 10.) Through D, the intersection of these bisectors, draw D E perpendicular to A C. (Problem 6.) With D as a center and radius D E, draw the required inscribed circle.

PROBLEM 18

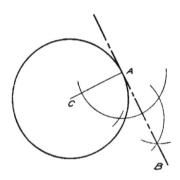

To draw a tangent to a circle at a given point on the circumference—T h r o u g h the given point A, draw the radial line A C. Erect a perpendicular to A C at the point A. (Problem 4.) The perpendicular A B is the required tangent at A, and the point A is the point of tangency.

PROBLEM 19

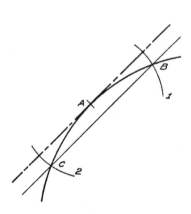

To draw a tangent to an arc at a point on the arc, when the center is not known—With the given point A as a center, draw arcs 1 and 2, cutting the arc at B and C. Draw the chord B C. Through the point A draw a line parallel to B C. (Problem 7.) This line will be tangent to the arc at the point A.

PROBLEM 20

To draw a tangent to a circle from a point outside the

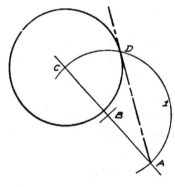

circle—From the given point A draw a line to the center of the circle. Find point B by bisecting the line A C. (Problem 1.) With B as a center and radius B A, draw semicircle 1, intersecting the circle at D. The line D A will be tangent to the circle at D.

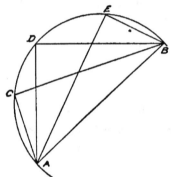

Note.—If from any point on the arc of a semicircle lines be drawn to the ends of the diameter, the included angle will be a right angle. In the figure, the angles A C B, A D B and A E B are right angles.

PROBLEM 21

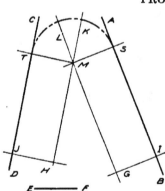

To draw an arc of a given radius tangent to two converging lines—Let A B and C D be the converging lines and E F the given radius. At any points, I and J, draw perpendiculars to A B and C D. (Problem 4.) Make I G and J H equal to E F. Draw H K and G L

parallel to the lines C D and A B. (Problem 7.)
From M draw M S and M T perpendicular to A B
and C D. (Problem 6.) With M as a center and
radius M T, draw the required arc tangent to the
converging lines at S and T.

PROBLEM 22

To draw an arc of a given
radius tangent to two circles
of fixed diameters—Let C
D be the given radius.
From A and B, the centers
of the given circles, draw
any indefinite lines A K
and B L. Make G K and
H L equal to C D. With
B as a center and radius
B L, draw arc 1. With
A as a center and radius
A K, draw arc 2 intersecting arc 1 at M. M is the
center of the required tangent arc. Lines M A and
M B will determine points
of tangency T and S.

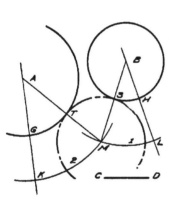

PROBLEM 23

To draw an arc of a given radius tangent to a given
line and arc—Let A B be the given radius, C D the
given line and E H S the given arc. Draw any radial

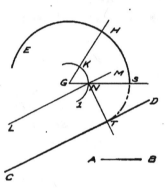

line G H.   Measure off H K equal to A B.   With G as a center and radius G K, draw arc 1.   Draw line L M parallel to and a distance equal to A B from C D.   (Problems 7 and 21.)   Where L M cuts arc 1, draw N T perpendicular to C D.   (Problem 6.)   With N as a center and radius N T, draw the required tangent arc.   T is one point of tangency, and the other one may be obtained by extending G N to S.

PROBLEM 24

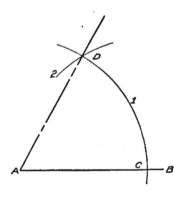

To construct an angle of 60 degrees—Let A B be one of the sides.   With A as a center and any radius, draw arc 1, cutting A B at C. With C as a center and the same radius, draw arc 2 cutting arc 1 at D.   Line A D will make an angle of 60 degrees with A B.

PROBLEM 25

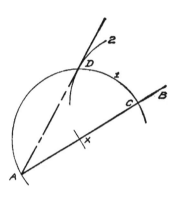

To construct **an angle of 30 degrees**—The line A B is one of the sides. With any point X as a center and radius X A, draw a semicircle cutting the line A B at C. With C as a center and the same radius, draw arc 2 cutting arc 1 at D. A D will make an angle of 30 degrees with A B.

PROBLEM 26

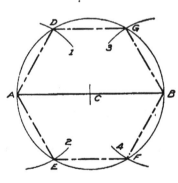

To draw **a hexagon, having given the long diameter** —Let A B be the long diameter. Find point C by bisecting A B. (Problem 1.) With C as a center and radius A C, draw a circle. With A as a center and the same radius, draw arcs 1 and 2, cutting the circle at D and E. With B as a center and the same radius, draw arcs 3 and 4, cutting the circle at G and F. Lines A D, D G, G B, B F, F E, and E A will form a hexagon.

PROBLEM 27

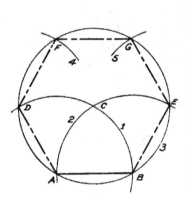

To draw a hexagon, when the length of one side is given—If A B is the given side, draw arcs 1 and 2 with A and B as centers and radius A B. With C as a center, draw circle 3 through A and B. With D as a center, and same radius, draw arc 4, cutting circle 3 at F. With E as a center and the same radius, draw arc 5 cutting circle at G. Lines A D, D F, F G, G E, and E B are the required lines of the hexagon.

PROBLEM 28

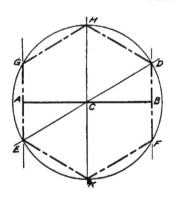

To draw a hexagon, the short diameter being given —Erect the perpendiculars G E and D F at the ends of the short diameter A B. (Problem 4.) Find point C by drawing the bisector H K. (Problem 1.) Make angle B C D equal 30 degrees. (Problem 25.) With

C as a center and a radius equal to the distance from C to D, draw a circle. Connecting points D H, H G, G E, E K, K F, and F D will give the required hexagon.

PROBLEM 29

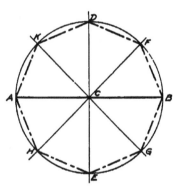

To draw an octagon, having given the long diameter —Find point C by bisecting the long diameter A B. (Problem 1.) Draw lines F H and K G making angles of 45 degrees with A B. (Bisect angles B C D and A C D.) With C as a center and a radius equal to the distance from C to B draw a circle. Connecting B F, F D, D K, K A, A H, H E, E G, and G B will give the required octagon.

CHAPTER XVII

GEOMETRICAL EXERCISES

These problems may be solved in a 5x7 rectangle. For the home work a cheap compass, made especially for the solution of problems in geometry and costing about 25 cents, may be employed. If these problems are inked, it is well to ink all given lines full, all results with a long and short dash, and to leave all construction lines in pencil. Show the construction for each problem entering into the solution of the exercise.

1. Draw an oblique line $3\frac{7}{16}''$ long and bisect it. (Problem 1, page 181.)

2. Bisect an arc of $3\frac{5}{16}''$ radius, having a chord of $4\frac{7}{16}''$. (Problem 2, page 181.)

3. Divide a horizontal line $4\frac{3}{16}''$ long into four equal parts. (Original.)

4. Divide the arc of a semicircle of $2\frac{7}{8}''$ radius into four equal parts. (Original.)

5. Locate two points $3\frac{15}{16}''$ apart. Draw an arc through these points with $3\frac{1}{4}''$ radius. (Original.)

6. Draw an oblique line $4\frac{1}{4}''$ long. Erect a perpendicular at a point on the line $2''$ from one end. (Problem 3, page 181.)

7. At a point $\frac{1}{2}''$ from the end of and on a horizontal line $5''$ long, erect a perpendicular to the line. (Problem 4, page 182.)

8. The two parallel sides of a trapezoid measure $2\frac{1}{8}''$ and $3\frac{5}{8}''$, respectively. These two sides are perpendicular to an oblique line $1\frac{7}{16}''$ long. Draw the trapezoid. (Original.)

9. From a point at least $2\frac{1}{2}''$ above and near the end of a horizontal line $4\frac{1}{2}''$ long, draw a perpendicular to the line. (Problem 5, page 182.)

10. Draw a perpendicular to a horizontal line which is $3\frac{7}{8}''$ long from a point at least $2\frac{1}{2}''$ from and over the center of the line. (Problem 6, page 183.)

11. Two lines $2\frac{3}{8}''$ and $3\frac{7}{8}''$ long, respectively, form a right angle. Draw the angle. (Original.)

12. Through a point not less than $1\frac{1}{2}''$ from an oblique line $3\frac{7}{8}''$ long draw a parallel to the line. (Problem 7, page 183.)

13. Draw a parallelogram in which two of the sides shall be $3\frac{7}{16}''$ long and $1\frac{5}{8}''$ apart. (Original.)

14. Divide a line $4\frac{1}{16}''$ long into three equal parts. (Problem 8, page 184.)

15. Draw two lines making any obtuse angle; copy the angle; have none of the lines horizontal and all at least $2\frac{1}{2}''$ long. (Problem 9, page 184.)

16. Draw any acute angle having sides at least $2\frac{5}{8}''$ long. Bisect the angle. (Problem 10, page 185.)

17. Bisect a right angle having sides at least $2\frac{15}{16}''$ long. (Original.)

18. Draw any obtuse angle with sides at least

$2\frac{1}{2}''$ long. Divide the angle into four equal parts. (Original.)

19. An oblique line $2\frac{11}{16}''$ long is the base of an equilateral triangle. Construct the triangle. (Problem 11, page 186.)

20. A vertical line $3\frac{1}{2}''$ long is one side of an equiangular triangle. Draw the triangle. (Original.)

21. Draw an isosceles triangle in which the base is a horizontal line $3\frac{9}{16}''$ long and the equal sides are $2\frac{5}{16}''$. (Problem 12, page 186.)

22. The altitude of an isosceles triangle is $3\frac{11}{16}''$ and the base is $2\frac{1}{8}''$. Draw the triangle. (Original.)

23. Draw an isosceles triangle in which the equal sides are $2\frac{3}{4}''$ long and form a right angle. (Original.)

24. Construct a scalene triangle having sides $2\frac{1}{2}''$, $3\frac{1}{2}''$ and $4\frac{1}{2}''$. (Problem 13, page 187.)

25. The base of a scalene triangle is $3\frac{7}{8}''$ long. One of the other sides makes a right angle with the base and the third side is $5\frac{1}{8}''$ long. Draw the triangle. (Original.)

26. The altitude of a triangle is $2\frac{3}{4}''$; the base is $3\frac{3}{8}''$ long; and one of the sides is $4\frac{13}{16}''$. Draw the triangle. (Original.)

27. Draw a square having sides $3\frac{3}{16}''$ long. (Problem 14, page 187.)

28. In a rectangle the parallel sides are $3\frac{7}{8}''$ and $2\frac{5}{8}''$, respectively. Draw the rectangle. (Original.)

29. The three sides of a triangle measure $4\frac{3}{8}''$, $3\frac{1}{8}''$ and $3\frac{1}{2}''$. Circumscribe a circle about this triangle. (Problem 15, page 188.)

30. Locate any three points and draw a circle through them. (Original.)

31. Circumscribe a circle about a right-angled, scalene triangle. The base of the triangle is $3\frac{1}{2}''$ and the altitude is $2''$. (Original.)

32. About a three inch square circumscribe a circle. (Problem 16, page 188.)

33. Within a circle of $3\frac{3}{8}''$ diameter draw a square with the corners touching the circumference of the circle. (Original.)

34. Inscribe a circle within a square having sides $3\frac{1}{2}''$ long. (Original.)

35. Inscribe a circle within a right triangle having sides bounding the right angle which measure $4\frac{1}{2}''$ and $3\frac{1}{2}''$. (Problem 17, page 188.)

36. At any point on the circumference of a circle $3\frac{1}{2}''$ in diameter, draw a tangent to the circle. (Problem 18, page 189.)

37. Without using the radius of the circle, draw a tangent to an arc of $3\frac{1}{2}''$ radius. (Problem 19, page 189.)

38. From a point $3\frac{3}{4}''$ from the center of a circle which is $3\frac{1}{4}''$ in diameter, draw a tangent to the circle. (Problem 20, page 189.)

39. Draw two tangents to an arc of $2\frac{3}{4}''$ radius from a point $4''$ from the center of the arc. (Original.)

40. Construct any right-angled triangle the longest side of which shall be $5''$ long. (See note following Problem 20, page 190.)

41. Draw two intersecting lines making any angle.

With a radius of $1\frac{1}{4}''$ draw an arc tangent to the two lines. (Problem 21, page 190.)

42. Two circles of $2\frac{1}{2}''$ and $3\frac{1}{2}''$ diameter have their centers $3\frac{5}{8}''$ apart. With a radius of $1\frac{1}{2}''$, draw an arc tangent to the two circles. (Problem 22, page 191.)

43. Draw two circles each having a radius of $1\frac{1}{16}''$ tangent to each other and also tangent to a circle $2\frac{1}{2}''$ in diameter. (Original.)

44. Having given three circles $2\frac{1}{4}''$, $2\frac{3}{4}''$ and $3''$ in diameter, draw them so that each shall be tangent to the other two. (Original.)

45. At a point $1\frac{1}{4}''$ from one end of a line which is $4\frac{1}{2}''$ long, erect a perpendicular. About a point on this perpendicular $2\frac{1}{4}''$ from the given line draw a circle of $1\frac{1}{8}''$ radius. With a radius of $1\frac{1}{4}''$ draw an arc tangent to the circle and the given straight line. (Problem 23, page 191.)

46. Using the circle and the straight line described in Exercise 45, draw an arc of $1''$ radius tangent to the circle and the straight line with its center outside of the given circle. (Original.)

47. Construct an angle of 60°, making the sides at least $2\frac{1}{4}''$ long. (Problem 24, page 192.)

48. A vertical line $2\frac{3}{8}''$ long forms one side of an angle of 30°. Construct the angle. (Problem 25, page 193.)

49. Draw an angle of 30°, without using the method described in Problem 25. (Combine Problem 24, page 192 and Problem 10, page 185.)

50. Construct an angle of 15° with one side a

horizontal line not less than 2½″ long. (Original.)

51. Draw an angle of 22½°. Make sides 2¼″ long. (Original.)

52. Construct an angle of 75°. (Original.)

53. Draw an angle of 37½°. (Original.)

54. The base of a right triangle is 3¾″ long and one of the angles is 30°. Draw the triangle. (Original.)

55. In an isosceles triangle the base measures 3½″ and the equal angles are 45°. Draw the triangle. (Original.)

56. The obtuse angle of a scalene triangle measures 135° and the base is 3″ long. Draw a triangle satisfying these requirements. (Original.)

57. The angle made by the equal sides of an isosceles triangle is 15° and the equal sides are 4″ long. Draw the triangle. (Original.)

58. The long diameter of a hexagon is 4″. Construct the hexagon. (Problem 26, page 193.)

59. Divide a circle of 1¾″ radius into six equal parts. (Original.)

60. Using a line 1½″ long as one side of a hexagon, construct the hexagon. (Problem 27, page 194.)

61. Draw a hexagon having a short diameter of 3⅜″. (Problem 28, page 194.)

62. Draw a polygon having a long diameter of 3⅞″ and twelve equal sides. (Original.)

63. Draw an octagon with the long diameter 4⅛″. (Problem 29, page 195.)

64. Inscribe a circle within an octagon whose circumscribing circle is 1¹⁵⁄₁₆″ radius. (Original.)